Natural Products

A Case-Based Approach for Health Care Professionals

Notices

The author, editors, and publisher have made every effort to ensure the accuracy and completeness of the information presented in this book. However, the author, editors, and publisher cannot be held responsible for the continued currency of the information, any inadvertent errors or omissions, or the application of this information. Therefore, the author, editors, and publisher shall have no liability to any person or entity with regard to claims, loss, or damage caused or alleged to be caused, directly or indirectly, by the use of information contained herein.

The inclusion in this book of any product in respect to which patent or trademark rights may exist shall not be deemed, and is not intended as, a grant of or authority to exercise any right or privilege protected by such patent or trademark. All such rights or trademarks are vested in the patent or trademark owner, and no other person may exercise the same without express permission, authority, or license secured from such patent or trademark owner.

The inclusion of a brand name does not mean the author, the editors, or the publisher has any particular knowledge that the brand listed has properties different from other brands of the same product, nor should its inclusion be interpreted as an endorsement by the author, the editors, or the publisher. Similarly, the fact that a particular brand has not been included does not indicate the product has been judged to be in any way unsatisfactory or unacceptable. Further, no official support or endorsement of this book by any federal or state agency or pharmaceutical company is intended or inferred.

Natural Products

A Case-Based Approach for Health Care Professionals

Karen Shapiro

Washington, D.C.

Managing Editor: Vicki Meade, Meade Communications
Acquiring Editor: Sandra J. Cannon
Copy Editor: L. Luan Corrigan, Corrigan Editorial Services
Proofreader: Amy Morgante
Indexer: Columbia Indexing Group
Book Design and Layout: Claire Purnell Graphic Design
Cover Design: Claire Purnell Graphic Design
Editorial Assistant: Kellie Burton

Photography:
All botanical photographs on the cover and inside this book, with the exception of Bitter Melon, are the property of Steven Foster Group, Inc., Eureka Springs, Arizona. ©2006 Steven Foster.
Bitter Melon by Jess Lopatynski; Brown Bag Medications by Patrick O'Connell.
All other photos by Karen Shapiro.
Plants shown on cover, clockwise from upper left: echinacea, soy, St. John's wort, and oats.

© 2006 by the American Pharmacists Association
Published by the American Pharmacists Association
2215 Constitution Avenue, N.W.
Washington, DC 20037-2985
www.aphanet.org

All rights reserved

APhA was founded in 1852 as the American Pharmaceutical Association.

No part of this book may be reproduced, stored in a retrieval system, or transmitted in any form or by any means, electronic, mechanical, photocopying, recording, or otherwise, without written permission from the publisher.

To comment on this book via e-mail, send your message to the publisher at aphabooks@aphanet.org

Library of Congress Cataloging-in-Publication Data
Shapiro, Karen, 1961-
    Natural products : a case-based approach for health care professionals / Karen Shapiro.
       p. ; cm.
    Includes bibliographical references and index.
    ISBN-13: 978-1-58212-069-0
    ISBN-10: 1-58212-069-2
    1. Evidence-based medicine. 2. Natural products. 3. Pharmacognosy.
4. Clinical medicine—Decision making. I. American Pharmacists Association. II. Title.
    [DNLM: 1. Biological Products—therapeutic use.   QW 800 S529n 2006]
    R723.7.S53 2006
    616—dc22
                                                                                                                2006001287

How to Order This Book
Online: www.pharmacist.com • By phone: 800-878-0729
VISA®, MasterCard®, and American Express® cards accepted.

Dedication

This book is dedicated to my father, Jerry Shapiro. His sudden death this past year has caused an almost unbearable pain to me, to my sister Ruth, and to his many fellow travelers. We miss his appreciation of life.

To see a World in a Grain of Sand
And a Heaven in a Wild Flower
Hold Infinity in the palm of your hand
And Eternity in an hour.

—William Blake, "Auguries of Innocence"

Table of Contents

Preface ix
Contributing Authors xi
Chapter 1: Introduction to Natural Products 1
Chapter 2: Dementia 15
Chapter 3: Migraine 25
Chapter 4: Osteoarthritis 37
Chapter 5: Erectile Dysfunction 51
Chapter 6: Benign Prostatic Hyperplasia 63
Chapter 7: Osteoporosis 71
Chapter 8: Menopause 85
Chapter 9: Depression 97
Chapter 10: Cholesterol Reduction 115
Chapter 11: Cardiovascular Disease 135
Chapter 12: Diabetes 159
Chapter 13: Cold and Flu 173
Chapter 14: Weight Loss 193
Chapter 15: Performance Enhancement 209
Index 229

Preface

I wrote this book to help schools and practitioners make a shift towards integrative medicine and to do my part to move medical practice from a disease-based to a healing-based approach.

Integrative medicine does not reject conventional medicine, nor does it accept unconventional practices without assessing them critically. It is a humanistic and fulfilling approach to medicine, and its benefits have been documented for both chronic and acute conditions. Natural products are one part of an integrative medicine approach.

Reading this book will give you a basic education in common natural products and the uses that are backed by solid evidence. Natural products can be helpful, harmful, or innocuous. I hope this text enables you to guide your patients so they use natural products in the most beneficial ways possible.

Karen Shapiro
Los Angeles, California
September 6, 2005

Contributing Authors

Jack J. Chen, PharmD, BCPS
Loma Linda University College of Pharmacy
Loma Linda, California

Eunice P. Chung, PharmD
Western University College of Pharmacy
Pomona, California

Christian R. Dolder, PharmD, BCPS
Wingate University School of Pharmacy
Wingate, North Carolina

Julia Ireland, DO
Private Practice
West Hollywood, California

Leila Khajehmolaei, PharmD
Los Angeles, California

James D. Scott, PharmD
Western University College of Pharmacy
Pomona, California

Karen Shapiro, PharmD, BCPS
Rancho Los Amigos National Rehabilitation Center
Los Angeles County Department of Health Services
Downey, California

Sandra Shibuyama, PharmD
St. Mary's Hospital
Long Beach, California

Winston Y. Wong, PharmD
Rancho Los Amigos National Rehabilitation Center
Los Angeles County Department of Health Services
Downey, California

CHAPTER 1
Introduction to Natural Products

Karen Shapiro

The dietary supplement business is a $12 billion industry in the United States.[1] Nearly 19% of Americans used dietary supplements in 2002, up from 14.2% in 1998–1999. Among people over age 65, use more than doubled during the same period.[2] Compared with 1994, when about 4000 dietary supplement products were on the market in the United States, there are now about 29,000, with many more added each year.[3] Although single-ingredient products are predominant, the trend today is toward multiagent supple-

KEY POINTS

- Natural products are regulated according to the Dietary Supplement Health and Education Act (DSHEA).
- The DSHEA allows dietary supplements to be marketed with little or no proof of safety or efficacy.
- The FDA must prove that a dietary supplement is harmful in order to have it removed from the market.
- A "natural" product may have been chemically altered in a laboratory.
- "Natural" does not necessarily mean safer.
- Consumers must consider safety, efficacy, and quality when selecting a product.
- Natural products should be tested by a reputable agency for content, disintegration, and contamination.
- Resources exist to help in the selection of a reputable product.
- A clinician should choose an easily updated reference source for natural products since information in this area changes rapidly.
- Many natural products have dose–response relationships and may have lag periods for full response.
- Patients should use an appropriate dose and trial period and monitor response to individual agents.
- Natural products can be subject to polypharmacy; it is usually best to use individual, rather than combination, products.
- Some natural products increase bleeding risk.
- Some natural products are enzyme inducers and can lower the concentration of certain drugs.
- Patients should be encouraged to bring all prescription and nonprescription products to medical appointments.
- All adverse reactions involving dietary supplements should be reported using the FDA's MedWatch program.
- Care of the environment and safe farming practices should be considered.

ments—formulas for joint health or cholesterol control, for example—and sales of these combination products are growing faster than sales of individual agents.[4]

Reflecting the rising interest in natural products to treat illness and promote wellness, this text is designed as a practical learning tool to familiarize users with key facts and walk them through decision-making processes regarding popular natural supplements on the market today.

Products are discussed within the framework of common health conditions, such as depression or diabetes, and patient case scenarios are included to illustrate key points and to help master important considerations. A case-based learning approach, common in pharmacology courses, is one of the best ways to understand and retain knowledge. This book is not designed to be an exhaustive reference; many compendia, comprehensive reviews, and databases on natural products already exist for that purpose. Recommendations for reference sources are included in this chapter.

Understanding how to safely integrate standard and alternative care in practice is necessary not only to benefit patients (for example, fish oils have anti-inflammatory properties), but also for safety reasons (the product may be harmful). Countless cases have been reported of interactions between drugs and natural products, and other safety issues have come to light, such as hepatoxicity, effects on laboratory tests, and product contamination. With so many people using natural products, often without the advice of a health care provider, competent clinicians need to know each product's benefits and drawbacks, backed by research.

TABLE 1-1
Top 10 Natural Products Used by Adults in the United States (2002)

RANK	PRODUCT
1	Echinacea
2	Ginseng
3	Ginkgo
4	Garlic
5	Glucosamine
6	St. John's wort
7	Peppermint
8	Fish oil
9	Ginger
10	Soy

Source: Reference 5.

NATURAL PRODUCTS DEFINED

What actually is a "natural product?" The Dietary Supplement Health and Education Act (DSHEA), passed in 1994, defined natural products as "dietary supplements" and defined the term "dietary supplement" as a product containing one or more of the following ingredients:

- Vitamin
- Mineral
- Herb or other botanical
- Amino acid
- Dietary substance for use by humans to supplement the diet by increasing the total daily intake
- A concentrate, metabolite, constituent, extract, or combination of the listed ingredients.

The product must be intended for ingestion in pill, capsule, tablet, or liquid form and cannot be marketed for use as a conventional food or as the sole component of a meal or diet. According to the DSHEA definition, which is followed in this book, natural products include herbals or plant products, such as feverfew and ginkgo biloba, and nonplant products, such as glucosamine, coenzyme Q10, creatine, and others.

Natural products are not regulated like conventional drugs. DSHEA allows dietary supplements to be marketed with little or no proof of safety or effectiveness. Supplements marketed before DSHEA was passed were "grandfathered" in and considered safe for continued consumer use based on history and experience. Dietary supplements marketed after 1994 must provide "reasonable assurance" of safety based on their presence in the food supply, history of use, or other evidence that supports reasonable safety expectations.

The burden of proof is on the Food and Drug Administration (FDA) to determine which products might be unsafe. In other words, marketers of dietary supplements do not need to conduct large-scale studies to determine if their products are safe. If the FDA is concerned about the safety of a product, the agency must prove that the product poses a threat in order to remove it from the market.

Data on efficacy are not required for dietary supplements to reach the market, and as a result, they cannot make claims to treat, cure, or diagnose disease. Thus, it is illegal for dietary supplements to claim to "treat cancer" or "treat pain from arthritis." Instead, supplements can only make unscientific structure/function claims, such as "supports healthy circulation" or "strengthens immune function." Unfortunately these claims are so nonspecific and nonscientific that it is possible they do more harm than good, allowing supplement makers to imply that products have curative or health properties, in some cases without evidence.

Evidence to Guide Decisions

Efficacy of natural products is widely variable, ranging from no benefit to mild or significant benefit. This book uses an evidence-based approach to help ensure that decisions are based on sound information. Providing solid guidance is not a simple task, however, especially if study data are preliminary, poor, or insufficient. Sometimes the data for the use of natural products are observational. For example, women who consume a diet rich in soy products have been found to report significantly fewer hot flashes and other vasomotor symptoms of menopause, an observation that has led to keen interest in using soy products for symptom relief. Yet observational data can be proven wrong with clinical trials.

The choice in using any therapeutic agent, in the end, is up to the patient. The role of a clinician is to help guide rational decision-making. When patients want to use a natural product, they should be told about data insufficiencies and safety concerns.

Keep in mind that even when a product has been proven to be useless, the placebo effect can contribute to benefit—especially when the patient or someone they trust is helping guide the therapy.[6]

What Does "Natural" Mean?

To consumers, the term "natural" often implies a safer product in comparison to conventional prescription medications, but "safer" is not always true. Natural products can be toxic and can be available in purified or manipulated forms as prescription drugs. Some chemotherapy drugs, including vincristine, vinblastine, and paclitaxel, as well as digoxin, colchicine, atropine, aspirin and some herbicides are plant-derived. Many plants produce chemicals whose purpose is to make sick or kill whatever bites into them. Major pharmaceutical companies have research teams devoted to finding new drugs from these often toxic (natural) plants in the rainforest. The bottom line is that safety depends on the particular product, not on whether it is termed "natural."

Everyone would likely agree that compounds found in the natural state and not structurally manipulated in a laboratory, such as plant extracts and desiccated plant products, are natural. What about a plant extract that has been manipulated in a laboratory to produce a different chemical compound? Pharmacologists refer to products that have been manipulated from raw materials as synthetic or semisynthetic.

Female hormone replacement therapy is a great example of the different ways in which terminology is used. Premarin®, the common form of prescription estrogen, is made from desiccated mare (female horse) urine. The hormones present in the horse urine are not altered in any way. Premarin could be considered to be a "natural" product.

The hormones in Premarin do not match female human sex hormones—as expected, given that they come from a horse. Many women consider this product "unnatural" (to them) and prefer to use "bioidentical" hormone replacement therapy (BHRT). The estradiol used in BHRT is identical to the estradiol produced by a pre- (not post-) menopausal woman. BHRT estradiol is derived from plant precursors, which are manipulated in the laboratory to match female hormones. To the pharmacologist who tweaked the estradiol from plant product precursors in a laboratory, the end-product is synthetic. The hormone replacement product Cenestin® is advertised as a natural, plant-derived alternative to Premarin. At the bottom of the same advertisements, in small print, it reads "synthetic conjugated estrogens." Is Premarin, Cenestin, or BHRT safer? Which offers more symptom relief? Only well-designed clinical trials will answer these questions.

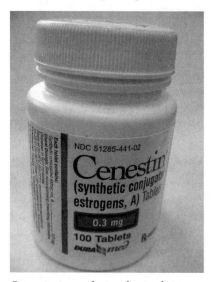

Cenestin is synthetic, despite being a plant-derived estrogen formulation

Glucosamine is a natural product used for osteoarthritis. The glucosamine present in joint supplement products is laboratory-derived from marine exoskeletons or is completely laboratory-manufactured. Glucosamine represents a natural product with good study data to support its use as a safer alternative to drugs such as ibuprofen, but its safety is not necessarily because it is "natural." In fact, there are "natural" nonsteroidal anti-inflammatory compounds similar to ibuprofen in food products in low doses, which could be concentrated and marketed—and would be expected to cause similar safety concerns as ibuprofen.[7]

Huperzine A, a compound derived from Chinese club moss, is advertised as a natural alternative to a class of drugs used to treat dementia, the acetylcholinesterase inhibitors. Yet the moss-derived compound has been manipulated structurally so that huperzine A is also an acetylcholinesterase inhibitor—not much different from prescription drugs in the same class, in either safety or effectiveness.

REPUTABLE RESOURCES

As in all therapeutics, competent clinicians know where to look when they need more information. Many textbooks are available that attempt to be inclusive and current, but in this rapidly changing field, they tend to be outdated by the time they come out in print. Rather than investing in natural product compendium texts, a better choice for clinicians who are grounded in the basics is to choose easily updated reference sources, such as those found online. Some key resources include the following:

- **The Natural Medicines Comprehensive Database** maintained by the Therapeutic Research Faculty is available in print, PDA, and online versions, and is the primary reputable source for natural product information. This resource provides clinically relevant information on more than 1000 products in an easy-to-use monograph format. References are linked for easy access to PubMed, and products are rated according to an evidence-based approach for safety and efficacy. Patient handouts and educational programs on hot topic areas are included. The online version is updated daily. The PDA version includes a checker for interactions between natural products and drugs. This resource is rated "superb" by a review in the *Journal of the American Medical Association*[8] and deemed "excellent" and "highly recommended" by the *Journal of the American Pharmacists Association.*[9] Web site: www.naturaldatabase.com.

- **The AltMedDex System,** part of the Micromedex Health Care Series Databases, is a reliable reference in a format many clinicians already use. Updated quarterly, AltMedDex covers dietary supplements and some form of alternative medicine. This service is available in many hospitals and clinics.

- **The Review of Natural Products,** published by Facts and Comparisons, is a well-referenced resource that includes more than 300 monographs describing clinical use, chemistry, pharmacology, interactions, toxicology, and important patient information. It includes most of the products patients are using. Web site: www.ovid.com.

- **The Natural Pharmacist Natural Medicine Encyclopedia,** by Steven Bratman and Richard Harkness, is a useful database of information on alternative medicine. It is not as thorough as the Natural Medicines Comprehensive Database but is easy to read and patient-friendly. An online version is free with a subscription to www.consumerlab.com (described below).

- **Office of Dietary Supplements, National Institutes of Health (NIH),** is a useful information source for patients and professionals. Fact sheets can be printed out on many common products and other helpful patient information is provided. Web site: http://ods.od.nih.gov.

- **The Cochrane Database of Systematic Reviews,** available online and on CD-ROM, provides succinct, well-researched summaries of many popular products. The database is published by the Cochrane Collaboration, a nonprofit resource for health care information known for rigorous quality standards. Web site: www.cochrane.org.

Choosing a Reputable Product

This book covers the first step in deciding which products to recommend (or not recommend) to patients. The next step is knowing how to pick the right bottle on the shelf.

Product quality varies significantly. The bottle the patient selects may contain little active ingredient—or it may contain even more than is listed on the label. Contaminants can be present. The tablet might not disintegrate. Products can be manufactured under Good Manufacturing Practices (GMPs)—or not.

The following organizations offer helpful information on dietary supplements as well as seals of approval indicating that some of the tablets of a product were tested and found to pass the agency's requirements. Although seals from these agencies are an advancement, they do not imply in any way that the product will benefit the patient's condition. That information needs to come from one of the previously mentioned references.

- **The ConsumerLab** Web site is a valuable resource for selecting an appropriate product. Using samples chosen at random, this organization tests for product purity, content, disintegration (will the tablet break down in the gut and be absorbed?), and for harmful levels of contaminants. The ConsumerLab seal on a bottle indicates a passing mark. The Web site, which provides test results of most common products, is available to patients and practitioners at a minimal cost. Web site: www.consumerlab.com.

- **The U.S. Pharmacopeia (USP),** a nonprofit organization that sets Federal standards for prescription drugs and dietary supplements, has a natural product approval system called the Dietary Supplement Verification Program (DSVP).[10] USP also has a testing program similar to ConsumerLab that allows the USP seal of approval on supplements that meet the testing requirements. Web site: www.usp.org.

- **NSF International,** a company that sets standards for quality in food, water, air, and consumer goods, has a certification program based on product tests and adherence to Good Manufacturing Practices. The NSF seal of approval appears on more than 60 brands of dietary supplements. Web site: www.nsf.org.

CARING FOR USERS OF NATURAL PRODUCTS

Taking a blanket approach in clinical practice and disavowing the use of all natural products—which clinicians tend to do when they do not understand these products' use and fear drug interactions—can be harmful to patients. Why? Because patients, who will likely use natural products anyway, will avoid discussing them to avoid a reprimand.

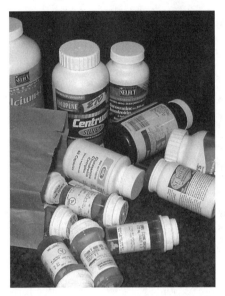

It's helpful to ask patients to bring all medicines to the initial appointment, including "whatever you take for your health and medical conditions." Any new products can be brought to subsequent appointments. Be sure to specify that all formulations should be included, such as teas, powders, liquids, and pills. If a patient's native language is not English, a better response may be obtained if the information is requested in his or her own language.

The possibility of interactions is an important argument for being aware of all products a patient is using. The most typical problem is the use of additive agents that increase bleeding risk, which can be an issue for patients using anticoagulants (e.g., warfarin) or antiplatelets (e.g., aspirin and clopidogrel) and for those at an increased bleeding risk for other reasons. Another significant concern is natural products that can lower the concentration of drugs (e.g., St. John's wort), resulting in therapeutic failure.

Drug interactions are not the only issue. If a product has the potential to cause liver toxicity, for example, appropriate monitoring is warranted, and if a product is harmful in other ways, there is an obligation to discourage its use. Safety considerations may be more important than concerns regarding efficacy.

Instruct patients to quantify results from a product as best they can by using diaries, pain scales, or other monitoring tools. Counsel them to stick with a reasonable dose over a trial period. Just like conventional agents, natural products have dose–response relationships and may have lag periods before a full response is achieved.

Polypharmacy, the use of multiple medications by a patient, often for the same condition, extends to the use of natural products. When patients take many compounds, it is hard to determine which individual agent is causing harm or benefit. Combination products often contain agents used for any similar condition and can contribute greatly to polypharmacy. For example, combination weight loss products usually include agents used for diabetes high cholesterol along with other agents with no known benefit for any condition. Of course, some ingredients added to these products may not be harmful, such as small amounts of vitamin C and calcium.

A BRIEF HISTORY OF NATURAL PRODUCTS USE: PAST TO PRESENT

For most of human history, resources for treating disease were limited. Plant remedies were used long before written history. Plant products still in use today have been found in Neanderthal burial sites from 60,000 years ago.[11]

In the written record, the use of plants as medicine dates back 5000 years to the Sumerians, who recorded plant prescriptions on clay tablets. Around the same period the Chinese Emperor Chi'en Nung recorded the important uses of more than 3000 medicinal plants. In the first century AD, the contributions of Greeks and Romans were recorded by the Roman scholar Pliny the Elder and by Dioscorides, a Greek army surgeon, who wrote *De Materia Medica*, the first compendium of herbal remedies in the West. This volume contains identification and instructions on the medicinal use of nearly 600 plants.

Up until the time of Paracelsus (1493–1541), a Swiss physician whose work encouraged the search for "active constituents" of medicinal plants, plant medicines involved the use of whole plant products. In fact, the word "drug" is derived from the Dutch droog, which means a dried plant substance. Paracelsus' work helped lay the foundation for chemistry and manufacturing.

In the early 1800s, morphine was isolated from opium and quinine was isolated from cinchona.[12,13] and the first proprietary drugs appeared in the 1890s.[14] These discoveries helped found the field of phyto, or plant, chemistry. Today approximately 25% of modern drugs contain one or more active ingredients originally derived from plants, although the plant source may since have been replaced.[15]

Throughout the 1800s, the science of pharmacology developed as understanding of disease processes and organic chemistry advanced. In

A Brief History of Natural Products Use continued

1820, the first *U.S. Pharmacopoeia* was published. It outlined the properties, dosage, dosage forms, purity standards, and production standards for the "drugs" of the day—many of which were herbal products. In 1828, salicin, a precursor to aspirin, was extracted from willow bark. And in 1852, aspirin was first synthesized in a laboratory. In 1846, diethylether ("ether") was first used in surgery as an anesthetic. Many other drugs were identified during the early 1800s, including chloral (used today as chloral hydrate) and chloroform.

In the 1860's, high casualties from the Civil War and widespread outbreaks of communicable disease raised the demand for large quantities of medicines, contributing significantly to the rise of mass pharmaceutical manufacturing.

By the early 1900s, medical schools were cutting back on the study of botany in favor of pharmacology. Penicillin was discovered in 1928 when Sir Alexander Fleming observed that colonies of the bacterium *Staphylococcus aureus* could be destroyed by the *Penicillium* mold, proving that antibacterial agents could kill certain types of disease-causing bacteria.

By the late 1940s penicillin was being mass-produced and none of the U.S. medical schools was teaching herbal medicine. The pharmaceutical industry had taken off and new generations of synthesized drugs were being discovered. Medicine in the United States and the modernized world had switched from natural-product-based treatments to synthesized drug treatments.

In the 1960s there was renewed interest in getting back to nature. By the 1990s, use of what is now called "alternative" or "complementary" or "integrative" medical approaches was skyrocketing, driven in large part by a rise in chronic diseases such as asthma and diabetes and dissatisfaction with visits to the doctor's office.

In a brief patient visit, laboratory values may be addressed but the patient may feel forgotten. Today's medical care, divided into specialties and removed from a holistic approach, can appear to the patient as compartmentalized, depersonalized care. Although many clinicians do their best to combat this trend, patients are looking for treatments that give them a sense of power and self-direction over their course of therapy. The movement toward complementary medicine has led patients to seek out natural products and alternative healing techniques.

Combination products should usually, but not always, be discouraged. Occasionally, combination products can be useful, such as those combining multivitamins and policosanol, a cholesterol-lowering agent. When patients need both, the combination product reduces the number of daily pills they must take and simplifies the medication regimen. It may also be less expensive than buying both products individually.

Very few natural products have been studied for use in pregnancy and lactation or in children. Unless there is significant reason to recommend a product, and the safety profile is established, avoid recommending natural products to these groups. There are exceptions, such as the use of riboflavin and magnesium for migraine prophylaxis in pregnancy. Use a reputable reference source to check for safety if there is any doubt.

All adverse reactions involving natural products should be reported to the FDA's MedWatch Program at www.fda.gov/medwatch or by calling 800-FDA-1088.

CARING FOR THE ENVIRONMENT

Unless a product can be manufactured ethically, its use should not be recommended. For example, cat's claw, a product used for many indications from asthma to HIV (with little evidence for efficacy) is harvested in the Peruvian rainforest. When natives harvest the vine, they replant and cultivate for the future, but opportunists ravage the vine and take no care for replacement. This product is not mass-produced and its use has caused damage to the rainforest.

In the 1960s, scientists discovered that an extract from the bark of the Pacific yew tree had anticancer properties. The active compound was identified as the chemotherapy drug paclitaxel. It took the bark of many of the slow-growing trees to treat one patient. Fortunately, a similar product was developed that can be derived semisynthetically from the needles of a relative of the Pacific yew.

The use of chemical fertilizers and pesticides, which introduce toxic compounds into the air, soil, and water—and ultimately the body—should be considered as well. Slow-growing plants, such as ginseng, can accumulate toxins over many years of growth.

One way to avoid the use of chemicals is to attempt to purchase organically grown plant products. These are not always available, but fortunately the number of reputable products that are organically grown is increasing.

SELF-ASSESSMENT

1. Which is TRUE concerning natural products:
   a. Medicinal plant products were found in human burial sites dated to 60,000 years ago.
   b. Dioscorides recorded the medicinal use of plants in his *De Materia Medica*.
   c. Many drugs today are plant-derived.
   d. All of the above.

2. The Dietary Supplement Health and Education Act was passed in:
   a. 1910
   b. 1920
   c. 1994
   d. 2004

3. Dietary supplements, similar to drugs, must be proven safe prior to marketing.
   a. True
   b. False

4. A mushroom mixture claims to "treat cancer" on the bottle label. This labeling is:
   a. Legal
   b. Illegal

5. A mixture of lipid-lowering products claims to "support healthy cholesterol levels" on the bottle. This labeling is:
   a. Legal
   b. Illegal

6. A manufacturer can market a product without studies demonstrating safety. Who is responsible for removing the product from the market if it proves to be unsafe?
   a. The manufacturer
   b. The NIH NCCAM
   c. The FDA
   d. None of the above

7. Natural products are often plant-derived. Care should be taken to protect the environment and to limit the use of harmful chemicals.
   a. True
   b. False

8. Bioidentical hormone replacement therapy (BHRT) is promoted as a natural way to treat menopausal symptoms. A patient using this therapy tells you that it is much more natural and safer than hormones made from a horse. Which of the following is correct?
   a. BHRT products contain only dessicated plant extracts.
   b. The hormones in BHRT are derived from plant precursors.
   c. The hormones in BHRT match those found in human females.
   d. b and c only

9. A patient with knee osteoarthritis wishes to try glucosamine. She is told by her clinician to go to the pharmacy and purchase the cheapest brand. This advice is:
   a. Reasonable; all products are likely to be similar in quality
   b. Not reasonable; a reputable product should be chosen

10. Seals of approval for product quality are issued by the following agencies:
    a. ConsumerLab
    b. USP
    c. NSF
    d. All of the above

Answers: 1-d; 2-c; 3-b; 4-b; 5-a; 6-c; 7-a; 8-d; 9-b; 10-d

REFERENCES

1. *Alternative Medicine.* Harvard Medical School, Palm Coast, FL: Harvard Health Publications; 2003.
2. Kelly JP, Kaufman DW, Kelly K, et al. Recent trends in use of herbal and other natural products. *Arch Intern Med.* 2005;165(3):281–6.
3. Atwater J, Montgomery-Salguero J, Roll DB. The USP Dietary Supplement Verification Program: helping pharmacists and consumers select dietary supplements. *US Pharm.* 2005;6:61–4.
4. Available at http://nccam.nih.gov. Accessed August 20, 2005.
5. Barnes P, Powell-Griner E, McFann K, et al. Advance Data Report 343. Complementary and alternative medicine use among adults: United States, 2002. May 27, 2004.
6. Gibbs WW. All in the mind. Fact or artifact? The placebo effect may be a little of both. *Sci Am.* 2001;285(4):16.
7. Beauchamp GK, Keast RS, Morel D, et al. Phytochemistry: ibuprofen-like activity in extra-virgin olive oil. *Nature.* 2005 Sep 1;437(7055):45–6.

8. Marty AT. Review, Natural Medicines Comprehensive Database. *JAMA.* 2000;283:2992–3.

9. Assessment of the quality of reference books on botanical dietary supplements. *J Am Pharm Assoc.* 2002;42(5):723–34.

10. Available at www.usp.org/USPVerified. Accessed August 20, 2005.

11. Rudgely R. *The Lost Civilizations of the Stone Age.* New York: The Free Press; 1999.

12. Cule J. *The Timetables of Medicine.* New York: Black Dog and Leventhal Publishers; 1999.

13. Jurna I. Sertürner and morphine—a historical vignette. *Schmerz.* 2003;17(4):280–3.

14. Haas LF. Pierre Joseph Pelletier (1788–1842) and Jean Bienaime Caventou (1795–1887). *J Neurol Neurosurg Psychiatry.* 1994;57(11):1333.

15. Nature's Pharmacy: Ancient Knowledge, Modern Medicine. Available at www.uihealthcare.com/depts/medmuseum/galleryexhibits/naturespharmacy/ naturepharmacy.html. Accessed January 11, 2005.

Chapter 2
Dementia

Karen Shapiro and Jack J. Chen

Alzheimer's disease is the most common form of dementia, affecting about 4.5 million men and women in the United States. The incidence of Alzheimer's disease increases with age, and it affects up to 50% of people older than 85 years of age. Since Alzheimer's is the most common form of dementia, natural products used for dementia are usually studied in Alzheimer's patients. However, ginkgo biloba, the most popular product used for dementia, has been studied in other types of dementia and in other types of cognitive dysfunction.

Gingko biloba

Alzheimer's disease requires a careful diagnosis prior to any treatment recommendation. Occasionally, dementia may have an identifiable cause and be treatable. Causes could include, among other things, a vitamin deficiency or severe depression. In most cases, the dementia cannot be cured and the best prescription agents offer only temporary improvement or a modest delay in symptom progression.

Most patients with moderate-to-severe Alzheimer's disease will likely be offered a

KEY POINTS

- Ginkgo biloba may provide modest benefit in dementia.
- Ginkgo biloba products should be standardized and tested for content.
- Ginkgo biloba extract inhibits platelet-activating factor and can increase bleeding risk.
- Ginkgo biloba supplements should be discontinued prior to elective surgery.
- Huperzine A has the same mechanism of action as the prescription acetylcholinesterase inhibitors.
- Vitamin E supplementation may be useful for slowing disease progression; however, the high doses used for dementia can contribute to an increase in all-cause mortality.

prescription drug. A few alternative agents in common use include ginkgo biloba, huperzine A, and vitamin E.

Ginkgo biloba

Ginkgo biloba is the best-selling plant medicine in Europe, where it is used for Alzheimer's disease, vascular dementia, peripheral claudication, and tinnitus. It is one of the top selling natural products in the United States. Ginkgo biloba extract is made from the leaves of the ginkgo tree, one of the oldest tree species on Earth. The leaves have two lobes, and thus the name biloba. Most of the ginkgo leaf used commercially comes from the Sumter plantation in South Carolina, where more than 12 million Ginkgo trees are cultivated.

Product	Dosage	Effect	Safety Concerns
Ginkgo biloba	120 to 240 mg daily, divided twice daily, in a product labeled to contain 24% flavone glycosides	Majority of studies support benefit for modest memory improvement	• Primary side effects are gastrointestinal complaints • Possibility of allergic skin reactions, including serious rash • Do not consume large amounts of ginkgo seed due to seizure risk; ginkgo supplements may decrease seizure threshold • Inhibits platelet-activating factor causing increased bleeding risk • Can decrease concentration of omeprazole and possibly other drugs
Huperzine A	100 to 200 mcg/day	May be as effective as the prescription acetylcholinesterase inhibitors—modest delay in symptom progression	• Gastrointestinal side effects • No additive benefit with prescription agents in same class—yet increased risk of toxicity
Vitamin E	1000 to 2000 IU/day	May slow rate of disease progression	• Do not recommend; dosage used for dementia unsafe

Ginkgo leaf extract contains many different flavonoids and terpenoids that have been studied individually and in combination. While some of the individual components may provide benefit, more benefit is derived from using the whole-leaf extract. Most products used in clinical studies have contained 24% to 25% flavone glycosides and 5% to 6% terpenoids (bilobalide and ginkgolides). Many products available to consumers do not contain these concentrations.[1]

The exact mechanism of action of ginkgo biloba extract is not well understood. The extract possesses anti-inflammatory properties and may help protect neurons from damage. Oxidative injury may develop secondary to

PATIENT CASE

MR. WALLACE

Mr. Wallace is a 72-year-old retired college professor who was diagnosed with Alzheimer's disease at a clinic visit. His wife had brought him to the doctor with concerns about her husband's memory. She stated that he frequently walked into rooms and forgot why he had gone there. She reported that his ability to recall things that had happened to him was diminished.

For example, his son and daughter-in-law had visited from the East coast, and a few days afterward he asked his wife when they were coming to visit. On one occasion he went to the store to pick up a few grocery items and returned 3 hours later, unable to account for why he was gone so long. He was prescribed donepezil 10 mg nightly. His other medications include warfarin for atrial fibrillation and benazepril for hypertension.

At his second clinic visit 10 months later, he is demonstrating significant decline in his cognitive function. At the initial appointment, when given four words, he was able to remember one after 5 minutes. Today, he cannot recall any of four given words and he does not recall being asked to memorize any words. At the previous visit, he was able to name a watch and pen, but he had some difficulty naming unusual objects—he called the watch buckle "a thing for closing it" and was unable to generate the name for a shoelace. Today, he could not name his watch and some other simple items.

His wife reports that he often forgets her name, which she finds particularly upsetting. She is concerned that his mood has worsened and he is sometimes irritable and angry. She states she is getting scared and is not sure if she will be able to care for him if his condition continues to worsen. She asks if ginkgo biloba, huperzine A, or vitamin E would help. She has been investigating Alzheimer's dementia on the Internet and is aware that people are using these products.

beta-amyloid-induced free radicals, inadequate energy supply, and inflammation. It is thought that when normal brain molecules are disrupted as a result of inflammation, amyloid beta proteins in the brain can misfold. Misfolded amyloid beta proteins are thought to have a critical role in the development of Alzheimer's dementia. The flavonoids have potent antioxidant and free radical scavenging properties.[2] Ginkgolides are selective antagonists of platelet aggregation induced by platelet-activating factor, an inflammatory mediator.[3-6]

The majority of evidence supports the use of ginkgo biloba as beneficial for modest memory improvement in patients with dementia.[7-9] One widely reported study in 1997 found no benefit; however, this study has been widely criticized for problems with study design.[10] Ginkgo biloba extract may not be as beneficial as acetylcholinesterase inhibitors, the typical first-line agents.[11] Many patients take both products together. A typical dose of ginkgo biloba for dementia is 120 to 240 mg/day, divided twice daily, in a tested product found to contain 24% flavone glycosides.

Safety Considerations/Drug Interactions

In the majority of patients, ginkgo biloba is well tolerated. The primary side effects are gastrointestinal complaints. A small number of patients report headache, dizziness, and allergic skin reactions.[12] Skin reactions, including serious rash, have occurred with the use of ginkgo taken orally. Gingko should not be used topically because it can irritate the skin. Ginkgo seeds contain a neurotoxin that can cause seizures if consumed in large amounts.[13] Ginkgo can lower the concentration of omeprazole, and possibly other proton pump inhibitors and other drugs.[14]

Due to platelet-activating factor inhibition, there have been numerous reports of ginkgo-associated bleeding.[15,16] Clinicians should consider whether concurrent use with warfarin, aspirin, or other antiplatelet agents and other natural products that increase bleeding is worth the risk. One study found no increase in the international normalized ratio (INR) in patients using both ginkgo and warfarin.[17] This means that the clinician may not be aware that the patient is at an increased bleeding risk. Ginkgo should be discontinued for at least 3 days and preferably longer (some clinicians recommend up to 2 weeks) prior to surgery.[18] Ginkgo has a short half-life, but the effect on platelet-activating factor is longer.

Huperzine A

Huperzine A is an acetylcholinesterase inhibitor derived from a particular type of Chinese club moss (*Huperzia serrata*) and then chemically purified.

Acetylcholinesterase inhibitors are an established therapy for Alzheimer's disease and dementia. The acetylcholinesterase inhibitors work by blocking the degradation of acetylcholine, a neurotransmitter that is important in learning and memory. Acetylcholine is greatly diminished in the brains of patients with Alzheimer's disease.

It is interesting that huperzine A is marketed as a natural product. It is made from a plant, but the plant extract then requires lab manipulation. By this definition, some of the chemotherapeutic drugs and many others (digoxin, colchicine, etc.) that are plant-derived would be considered "natural." This emphasizes the fact that natural may or may not mean safer. In this case, the product is relatively safe, but whether it is "natural" depends on the user's definition.

Huperzine A

The available data suggest that huperzine A may be as effective as the prescription acetylcholinesterase inhibitors.[19,20] It may also be neuroprotective.[21] The duration of action and enzyme specificity is longer than tacrine, an older acetylcholinesterase inhibitor.[22] Typical doses used for dementia are 100 to 200 mcg/day.

SAFETY CONSIDERATIONS/DRUG INTERACTIONS

Huperzine A, similar to other acetylcholinesterase inhibitors, can cause gastrointestinal upset. Much higher doses than those recommended here could theoretically cause cholinergic toxicity. Symptoms of cholinergic toxicity include bradycardia, bronchial hypersecretion and bronchoconstriction, skeletal muscle fasciculation and twitching, ataxia, and seizures. Huperzine A should not be taken concurrently with the prescription acetylcholinesterase inhibitors.

VITAMIN E

Vitamin E (α-tocopherol) is an antioxidant that prevents cellular damage. Free radicals contain an unpaired electron that can cause oxidative damage to cells and result in cell death. Antioxidants, such as vitamin E, prevent this damage by binding to the free radical and neutralizing the unpaired electron. Vitamin E is present in many oils, grains, nuts, and fruit. The Recommended Daily Intake (RDI) for vitamin E is 15 mg (equivalent to 22 vitamin E IUs of natural α-tocopherol or 33 vitamin E IUs of synthetic vitamin E) and is easily obtained from the diet.

A randomized, blinded trial with 341 subjects compared vitamin E 2000 IU/day to selegiline or placebo. Subjects taking vitamin E had a slower rate of disease progression (by 200 days) than placebo, and similar to the delay seen with subjects taking selegiline. Death was not delayed in either group.[23] A retrospective chart review of 130 patients found that vitamin E supplementation of at least 1000 IU/day might slow disease progression.[24] The data are not conclusive, but appeared promising until more recent results concerning vitamin E safety became available (see Safety Considerations/Drug Interactions). The studies for dementia used doses of 1000 to 2000 IU/day.

Safety Considerations/Drug Interactions

Doses of vitamin E for dementia are much higher than the RDI. Previously, high doses of vitamin E were thought to be innocuous, but this is no longer true. A meta-analysis involving 135,967 participants in 19 clinical trials studying the use of vitamin E found a statistically significant relationship between vitamin E dosage and all-cause mortality, with increased risk of dosages greater than 150 IU/day. The study concluded that doses greater than 400 IU/day should be avoided.[25,26]

Patient complaints from vitamin E supplementation are uncommon, except for the occasional gastrointestinal upset. High doses of vitamin E can elevate the INR and should be used with caution, if at all, in patients taking warfarin or antiplatelet agents.

Patient Discussion

Mr. Wallace is demonstrating more advanced disease and may be a candidate for a drug such as memantine, an N-methyl-d-asparate- (NMDA-) receptor antagonist. Alzheimer's disease is progressive and his condition will eventually worsen. The addition of a natural product might provide some mild benefit, but not to the extent that his wife requires. At this point, the caregiver (the wife) requires assistance with her husband's care.

Ginkgo biloba is the best-studied product and can be recommended for use in most patients. This patient is taking warfarin and the clinician must consider the increased risk of bleeding prior to recommending use. Although the INR may not be elevated secondary to ginkgo use, the risk for bleeding, due to platelet-activating factor inhibition, will be elevated.

Huperzine A should not be recommended for this patient, who is already using a drug, donepezil, with the same mechanism of action. Vitamin E, at high doses, may be detrimental and should not be recommended at this time.

Self-Assessment

1. Which is TRUE concerning ginkgo biloba:
   a. The tree is one of the oldest living tree species on Earth.
   b. Components of the plant extract are thought to work together to provide benefit.
   c. Ginkgo biloba extract has anti-inflammatory properties.
   d. All of the above.

2. Which of the following agents inhibits platelet-activating factor?
   a. Ginkgo biloba
   b. Huperzine A
   c. Vitamin E
   d. None of the above

3. Which is the correct dose of ginkgo to recommend for dementia?
   a. 1 to 2 mg/day
   b. 120 to 240 mg/day
   c. 0.2 mcg/day
   d. None of the above

4. Which is TRUE concerning ginkgo biloba:
   a. The seeds contain a neurotoxin which can cause seizures.
   b. Chewing the seeds is the best way to treat dementia.
   c. Ginkgo can be recommended as a topical agent.
   d. Ginkgo poses no risk for serious skin rashes.

5. Which is TRUE concerning huperzine A:
   a. It can be recommended for use in patients taking donepezil.
   b. The mechanism of action involves blocking the dopamine receptor.
   c. Gastrointestinal upset is possible.
   d. Cholinergic toxicity is likely with recommended doses.

6. Which is TRUE concerning vitamin E:
   a. Rich food sources include many oils, grains, nuts, and fruit.
   b. Most people get adequate vitamin E through the diet.
   c. Vitamin E is a potent antioxidant.
   d. All of the above.

7. Which is TRUE concerning vitamin E:
   a. The RDI is 15 mg.
   b. 15 mg is equivalent to 22 IU of natural a-tocopherols.
   c. 15 mg is equivalent to 2000 IU of natural a-tocopherols.
   d. A and B

8. Which is TRUE concerning vitamin E:
   a. Doses of 2000 IU/day are considered safe for most patients.
   b. There is an increased risk of all-cause mortality at doses greater than 150 IU/day.
   c. Vitamin E, at recommended doses for dementia, can cause significant diarrhea and flatulence.
   d. Vitamin E, at any dose, has no effect on the INR.

9. A patient wishes to give a natural product to her mother who has been diagnosed with Alzheimer's dementia. Her daughter did not want to give her any more prescription drugs, because she believes they are all toxic and states her Mom is taking too many already. Her mother takes clopidogrel (an antiplatelet agent), amlodipine, simvastatin, and aspirin. The best product to recommend is:
   a. Ginkgo biloba
   b. Huperzine A
   c. Vitamin E
   d. None of the above

10. A patient is using donepezil. It is reasonable to recommend huperzine A concurrently.
    a. True
    b. False

Answers: 1-d; 2-a; 3-b; 4-a; 5-c; 6-d; 7-d; 8-b; 9-b; 10-b

REFERENCES

1. Product review. Ginkgo biloba. Available at www.ConsumerLab.com. Accessed August 16, 2005.

2. Ilieva I, Ohgami K, Shiratori K, et al. The effects of Ginkgo biloba extract on lipopolysaccharide-induced inflammation in vitro and in vivo. *Exp Eye Res.* 2004 Aug;79(2):181–7.

3. Ho GJ, Drego R, Hakimian E, et al. Mechanisms of cell signaling and inflammation in Alzheimer's disease. *Curr Drug Targets Inflamm Allergy.* 2005 Apr;4(2):247–56.

4. Pasinetti GM. From epidemiology to therapeutic trials with anti-inflammatory drugs in Alzheimer's disease: the role of NSAIDs and cyclooxygenase in beta-amyloidosis and clinical dementia. *J Alzheimers Dis.* 2002 Oct;4(5):435–45.

5. Cole GM, Morihara T, Lim GP, et al. NSAID and antioxidant prevention of Alzheimer's disease: lessons from in vitro and animal models. *Ann N Y Acad Sci.* 2004 Dec;1035:68–84.

6. Schmidt R, Schmidt H, Curb JD, et al. Early inflammation and dementia: a 25-year follow-up of the Honolulu–Asia Aging Study. *Ann Neurol.* 2002 Aug;52(2):168–74.

7. Hopfenmuller W. [Evidence for a therapeutic effect of Ginkgo biloba special extract. Meta-analysis of 11 clinical studies in patients with cerebrovascular insufficiency in old age]. German. *Arzneimittelforschung.* 1994;44:1005–13.

8. Kanowski S, Herrmann WM, Stephan K, et al. Proof of efficacy of the Ginkgo biloba special extract EGb 761 in outpatients suffering from mild to moderate primary degenerative dementia of the Alzheimer type or multi-infarct dementia. *Pharmacopsychiatry.* 1996;29:47,56.

9. Kleijnen J, Knipschild P. Ginkgo biloba. *Lancet.* 1992;340:1136–9.

10. Le Bars PL, Katz MM, Berman N, et al. A placebo-controlled, double-blind, randomized trial of an extract of Ginkgo biloba for dementia. North American EGb Study Group. *JAMA.* 1997;278:1327–32.

11. Wettstein A. Cholinesterase inhibitors and Ginkgo extracts—are they comparable in the treatment of dementia? Comparison of published, placebo-controlled efficacy studies of at least six months duration. *Phytomedicine.* 2000;6:393–401.

12. DeFeudis FV. *Ginkgo biloba Extract (EGb 761): Pharmacological Activities and Clinical Applications.* Paris, France: Elsevier Science; 1991:143,146.

13. Arenz A, Klein M, Fiehe K, et al. Occurrence of neurotoxic 4'-O-methylpyridoxine in ginkgo biloba leaves, ginkgo medications and Japanese ginkgo food. *Planta Med.* 1996;62:548–51.

14. Yin OQ, Tomlinson B, Waye MM, et al. Pharmacogenetics and herb–drug interactions: experience with Ginkgo biloba and omeprazole. *Pharmacogenetics.* 2004;14:841–50.

15. Fong KC, Kinnear PE. Retrobulbar haemorrhage associated with chronic Gingko biloba ingestion. *Postgrad Med J.* 2003;79:531–2.

16. Hauser D, Gayowski T, Singh N. Bleeding complications precipitated by unrecognized Gingko biloba use after liver transplantation. *Transpl Int.* 2002;15:377–9. Epub 2002 Jun 19.

17. Engelsen J, Dalsgaard N, Winther K. The health care products Coenzyme Q10 and Ginkgo biloba do not interact with warfarin. *Thromb Haemost.* 2001(Suppl);Abstract No. P796.

18. Ang-Lee MK, Moss J, Yuan CS. Herbal medicines and perioperative care. *JAMA.* 2001;286:208–16.

19. Zhang SL. [Therapeutic effects of huperzine A on the aged with memory impairment]. Chinese. *New Drugs and Clinical Remedies.* 1986;5:260–2.

20. Zhang RW, Tang XC, Han YY, et al. Drug evaluation of huperzine A in the treatment of senile memory disorders. *Zhongguo Yao Li Xue Bao*. 1991;12:250–52. Chinese; English abstract.

21. Wang R, Tang XC. Neuroprotective effects of huperzine A. A natural cholinesterase inhibitor for the treatment of Alzheimer's disease. *Neurosignals*. 2005;14(1–2):71–82.

22. Wang H, Tang XC. Anticholinesterase effects of huperzine A, E2020, and tacrine in rats. *Zhongguo Yao Li Xue Bao*. 1998;19:27–30.

23. Sano M, Ernesto C, Thomas RG, et al. A controlled trial of selegiline, alpha-tocopherol, or both as treatment for Alzheimer's disease. The Alzheimer's Disease Cooperative Study. *N Engl J Med*. 1997;336:1216–22.

24. Klatte ET, Scharre DW, Nagaraja HN, et al. Combination therapy of donepezil and vitamin E in Alzheimer disease. *Alzheimer Dis Assoc Disord*. 2003;17:113–6.

25. Lonn E, Bosch J, Yusuf S, et al. HOPE and HOPE-TOO Trial Investigators. Effects of long-term vitamin E supplementation on cardiovascular events and cancer: a randomized controlled trial. *JAMA*. 2005;293:1338–47.

26. Miller ER 3rd, Pastor-Barriuso R, Dalal D, et al. Meta-analysis: High-dosage vitamin E supplementation may increase all-cause mortality. *Ann Intern Med*. 2005;142:37–46.

Chapter 3
Migraine

Karen Shapiro and Jack J. Chen

The World Health Organization now recognizes migraine as one of the most disabling medical conditions. There are 50 million Americans who suffer from chronic, severe headaches. Migraine is the most common type and is often misdiagnosed. Patients with migraine report an average of one to five attacks per month of moderate-to-severe pain, usually unilateral and accompanied by other symptoms, such as gastrointestinal upset, photophobia, and phonophobia.[1] Many patients would benefit from a combination of acute therapy (to treat attacks when they occur) and prophylactic therapy (to reduce the frequency of attacks). The natural products that are efficacious in treating an acute attack are caffeine, available in over-the-counter migraine products such as Excedrin® migraine, and magnesium, which is occasionally used in the acute care setting. Triptan agents, along with other classes of drugs, are used to resolve the acute attack. Common natural products used for prophylaxis include feverfew, butterbur, riboflavin, coenzyme Q10 (coQ10), and magnesium.

KEY POINTS

- Studies of feverfew in migraine prophylaxis offer conflicting results, which may be attributable to poor product quality.
- Preliminary data support the use of butterbur in migraine prophylaxis.
- Riboflavin is another name for vitamin B2, an essential nutrient in energy production.
- Riboflavin, in relatively high doses, may be useful for migraine prophylaxis.
- Riboflavin is inexpensive and well tolerated.
- Coenzyme Q10 may be useful for migraine prophylaxis.
- Magnesium is being studied for migraine prophylaxis and may provide benefit.
- Magnesium, in the recommended doses, is inexpensive and is well tolerated with the exception of mild gastrointestinal upset and loose stools in 20% of patients.
- Feverfew, butterbur, riboflavin, coenzyme Q10, and magnesium require larger, well-designed clinical trials in order to establish efficacy for migraine prophylaxis.

Feverfew

Feverfew (*Tanacetum parthenium*) is a perennial herb native to Southeastern Europe. According to Plutarch, the herb was named parthenium because it supposedly saved the life of a masonworker who fell from the roof of the Parthenon as it was being built. The plant is also known as midsummer daisy. This pretty plant has flowers with yellow center disks and white petals. The aromatic leaves have been used traditionally for inflammatory conditions

Product	Dosage	Effect	Safety Concerns
Feverfew	80 to 100 mg/day	Conflicting study data; may decrease frequency by 15% to 24%; 4 to 8 weeks for benefit	• May have additive risk with antiplatelets and anticoagulants • May decrease concentration of some medications • Avoid chewing leaves • Do not discontinue abruptly • Do not use in pregnancy or lactation
Butterbur	75 mg twice daily	May decrease frequency by 48% to 60%; may take 16 weeks for maximum benefit	• Use pyrrolizidine-free extract • Avoid use in liver disease
Riboflavin	400 mg/day	May decrease frequency by > 50%; allow at least 4 weeks for benefit	• Safe at recommended doses, including in pregnancy
Coenzyme Q10	100 mg three times daily	May decrease frequency by 46% to 50%; allow at least 4 weeks for benefit	• Use caution with certain drugs; see text
Magnesium	600 mg/day	May decrease frequency by ~40%; allow at least 4 weeks for benefit	• Safe at recommended doses, including in pregnancy • Can cause loose stools

and headaches. Currently, the most common use of feverfew is migraine prophylaxis.

The mechanism of action for feverfew in migraine prophylaxis remains unclear. Parthenolide, the principal sesquiterpene lactone, has been studied extensively as the proposed active ingredient. However, one study looking at a concentrated extract of parthenolide found no benefit for prevention of migraine.[2] It seems more likely that some combination of feverfew constituents is contributing to an antimigraine effect. This effect may be due to an inhibition of serotonin release, prostaglandin synthesis, and platelet aggregation.[3,4] Feverfew components might also prevent vascular muscle contraction.[5] Parthenolide may, however, be useful as an anti-inflammatory and chemotherapeutic agent.[6,7]

Feverfew

Studies of feverfew in migraine prophylaxis offer conflicting results. Two randomized, placebo-controlled trials that used whole feverfew leaf found feverfew effective as a prophylactic agent. The onset of benefit was observed in 4 to 8 weeks.[8,9] Others found no benefit.[10,11] One of the two trials that found

PATIENT CASE

Carol

Carol is a 28-year-old, 70-kg white female who complains of a severe, right-sided throbbing headache for the past 8 hours. The episode started with blind spots in both eyes, flashing lights, photophobia, nausea, vomiting, and a tingling feeling in the fingers of her left hand. These symptoms gradually subsided as the headache started, although the nausea persists. Carol usually uses sumatriptan for acute pain relief but ran out. She states this medication works fine and asks for a refill prescription. She uses the sumatriptan four or five times monthly.

Her doctor has been trying to find a daily medication she can tolerate to help reduce the headache frequency. She has tried propranolol, which made her tired, and divalproex sodium, which made her gain weight. She asks if there is a natural product that does not have side effects. She has heard about feverfew and asks if it works. "I'm desperate," she says, "to try something else. These headaches are killing me!"

efficacy reported a reduction in migraine frequency of 24%. In trials with this product and the other products presented here, a placebo reduces migraine frequency by approximately 15%, which makes the overall benefit for the active product less than 10%. Other products discussed later provide a greater reduction in migraine frequency. With feverfew, some of the variability in study results appears to be due to product variability. The only way to determine the actual benefit is to have well-designed trial data, which at present are not available.

Patients should be counseled to choose whole-leaf, encapsulated products, in dosages of 80 to 100 mg per day.

Safety Considerations/Drug Interactions

Feverfew is tolerated better than the available prescription agents. There is a theoretical risk of additive antiplatelet or anticoagulant effects when using feverfew with drugs such as warfarin, aspirin, and clopidogrel. Patients using these combinations should be aware of a potential increased risk of bleeding and understand how to monitor for bleeding. There are preliminary data indicating that feverfew may inhibit hepatic enzymes, which would increase the concentration of substrate drugs. Chewing feverfew leaves can cause oral ulcers.[10] Feverfew products may contain little or no dried leaf.[12] Upon abrupt discontinuation of feverfew, patients may experience a "postfeverfew syndrome" characterized by anxiety, insomnia, joint stiffness, and rebound headaches. Feverfew should be not be taken by women who are pregnant or breastfeeding.

Choosing a reputable product is essential because product quality is variable. Feverfew is a member of the ragweed plant family and patients with allergies to this plant family (including chrysanthemums, marigolds, and daisies) should avoid this product.

Butterbur

Butterbur grows in wet, marshy areas and along rivers in North America, parts of Asia, and Europe. It resembles rhubarb and is also called bog rhubarb. The leaves, rhizomes, and roots of butterbur contain the sesquiterpene compounds petasin and isopetasin.[5] These constituents inhibit leukotriene synthesis, which may also contribute to butterbur's antispasmodic and anti-inflammatory actions.[13,14]

Butterbur

Butterbur extracts contain other components that may contribute to efficacy, including flavonoids, tannins, and pyrrolizidine alkaloids.[15]

A standardized extract of butterbur has been used in randomized, placebo-controlled, double-blind clinical trials (Petadolex®, Weber & Weber International GmbH & Co., Germany.) This formulation reduced the frequency of migraines by 48% to 60%, along with a reduction in the intensity of existing migraines.[16,17] Butterbur also appears effective for allergic rhinitis.[18]

The benefit from this product increases with time. In one of the trials referenced previously, maximum benefit was not seen until the 16th week. A typical dose is 75 mg twice daily.[17]

Safety Considerations/Drug Interactions

Butterbur leaf and rhizome contain pyrrolizidine alkaloids, which are hepatotoxic and potentially carcinogenic.[19] Fortunately, these compounds can be safely removed. Consumers need to choose pyrrolizidine-free extracts. It is prudent to avoid all butterbur products, including pyrrolizidine-free formulations, in persons with liver disease. Side effects from the use of pyrrolizidine-free extracts are rare and consist primarily of mild gastrointestinal complaints.

Riboflavin

Riboflavin (vitamin B2) is an essential nutrient critical to the body's production of adenosine triphosphate (ATP), the main cellular energy source. Riboflavin is the precursor of flavin mononucleotide and flavin adenine dinucleotide, which are required for the activity of flavoenzymes involved in the electron transport chain. Riboflavin supplementation is thought to remedy a type of mitochondrial dysfunction which results in impaired oxygen metabolism. The impaired oxygen metabolism may be contributing to an increase in migraine frequency.

Riboflavin 400 mg was compared to placebo in 55 chronic migraine patients in a randomized, 3-month trial. Riboflavin was significantly superior to placebo in reducing the attack frequency (p = 0.005), headache days (p = 0.012), and migraine index (p = 0.012). Most of the participants in the treatment arm experienced more than a 50% reduction in migraine frequency. No serious adverse events were reported.[20] A second open-label study showed similar benefit.[21] Larger studies are required to provide a more definitive conclusion regarding benefit.

The Recommended Dietary Intake (RDI) for riboflavin is 1.1 mg/day for women and 1.3 mg/day for men.[22] Riboflavin deficiency is rare, except in alcoholics. The dosage typically used for migraine prophylaxis is 400 mg/day. A trial period should extend for at least 4 weeks. This dose is high compared to the RDI, but it is a safe daily intake.

Safety Considerations/Drug Interactions

No toxic effects of riboflavin at doses of 400 mg/day have been reported. At present, the low cost (less than 30 cents/day) and tolerability of riboflavin suggest that it could be tried as a reasonable option. It is reasonable to recommend in pregnancy.

Coenzyme Q10 (CoQ10)

CoQ10, also known as ubiquinone, is a naturally occurring antioxidant compound. The name of this supplement comes from the word ubiquitous, which means "found everywhere." Indeed, CoQ10 is found in every cell in the body. It plays a fundamental role in the mitochondria in energy production. The body normally produces sufficient CoQ10, although some medications such as statins may interfere with this process. CoQ10 levels in the body decline with age and certain disease states, including heart disease.

CoQ10 is used most commonly for congestive heart failure; however, the data for heart failure use are inconclusive. Data for migraine prophylaxis appear better, but the two available trials are small and only one is blinded and placebo-controlled. The other is open-label design.[23,24] In the open-label trial, mean reduction in migraine frequency was 55% at the end of 3 months. In the second trial, the reduction in migraine frequency was 46%. The dose used in the blinded, controlled trial was 100 mg three times daily and is the dose typically recommended for this indication.

Safety Considerations/Drug Interactions

CoQ10 does not cause significant adverse events. Mild gastrointestinal events occurred in clinical trials, with similar frequency to the placebo groups. Concomitant administration of CoQ10 and the chemotherapy drug doxorubicin (Adriamycin®) should be avoided as CoQ10 can alter the metabolism of doxorubicin and increase the concentration of a potentially toxic metabolite. CoQ10 therapy, however, has been used to help prevent cardiac toxicities of doxorubicin, when used after the cessation of chemotherapy.

Cholesterol-lowering drugs such as simvastatin (Zocor®), lovastatin (Mevacor®) and gemfibrozil (Lopid®) may decrease plasma and tissue CoQ10 levels. It is unclear whether normalizing coenzyme Q10 levels via supplementation will benefit patients taking these statins or gemfibrozil. Pravastatin (Pravachol®) and atorvastatin (Lipitor®) do not lower coenzyme Q10 levels.[25] The beta-blockers propranolol and metoprolol may also inhibit coenzyme Q10-dependent enzymes and ultimately lower CoQ10 levels.

CoQ10 is structurally similar to vitamin K. Therefore, a procoagulant effect (resulting in a decreased international normalized ratio) when combined with warfarin has been suggested, but a small study found no interaction between CoQ10 and warfarin.[26]

MAGNESIUM

Magnesium is an essential nutrient found in significant quantities throughout the body and used for numerous purposes, including muscle relaxation, blood clotting, and the manufacture of ATP. In the ambulatory population, magnesium is used most commonly as a laxative and antacid. It is also used as a "natural" calcium channel blocker for hypertension.

Most people get adequate magnesium through the diet, although deficiencies in certain disease groups, including hypertension, diabetes, and migraine, may be underrecognized.[27,28] The RDI is 420 mg/day for males and 320 mg/day for females.[22]

In one double-blind study of 81 patients, supplementation with magnesium 600 mg/day reduced migraine frequency by 42%, compared to 16% in the placebo group.[29] Twenty percent of subjects using magnesium reported diarrhea and, less often, gastrointestinal upset. Two other smaller studies showed benefit. The first study involved 43 patients using magnesium 600 mg/day in a double-blind, crossover pilot study.[30] Similar to other studies, onset for effectiveness was 4 weeks. The second double-blind study involved 20 women with premenstrual migraine taking magnesium 360 mg/day. In patients taking magnesium, there was a decrease in the number of days with headache and a significant decrease in pain.[28] A reasonable magnesium dose to recommend is 600 mg/day, for at least a 4-week trial. All magnesium salts are absorbed fairly well, although the most common magnesium formulation, magnesium oxide, is thought to be less well absorbed than magnesium citrate.

SAFETY CONSIDERATIONS/DRUG INTERACTIONS

Magnesium can cause diarrhea (more accurately, loose stools), and to a lesser extent, gastrointestinal upset, particularly at these higher doses. With a larger daily intake it is reasonable to consider the risk of hypermagnesemia, which can produce fatal arrhythmias. Known cases of hypermagnesemia involved ingestion of much larger amounts than those proposed here, such as ingestion of large quantities of epsom salts or excessive quantities used in very young children.[31,32] In patients with significantly reduced renal function, this may be a more significant concern due to reduced elimination. Patients with heart disease should not take excessive doses of magnesium without consulting their physician due to a preexisting risk for arrhythmias. Quinolone antibiotics, such as ciprofloxacin, and tetracycline antibiotics should be taken

2 hours before or 4 hours after magnesium supplements. Patients need to choose a formulation that has been tested for lead content. Magnesium occurs naturally with small amounts of lead; however, some supplements contain unsafe amounts.[33] Magnesium, at this suggested dose, is safe in pregnancy.

PATIENT DISCUSSION

Prior to discussing an appropriate natural product for Carol, it should be noted that her prescription options for prophylaxis are not yet exhausted. She might find that other prescription agents are better tolerated. She may also benefit from a combination of medication and a nonpharmacologic intervention for migraine prophylaxis. Biofeedback and relaxation therapy both reduce the frequency of migraines by about 50%, which is roughly equivalent to propranolol and other prophylactic therapies.[34]

She should be counseled to record a migraine history and attempt to identify triggers, in the hope that the triggers could be avoided. Triggers include hormonal changes; weather patterns; bright, flashing, or fluorescent lights; foods such as chocolate, wine, cheese, and caffeine; aspartame; and patient-specific food sensitivities.[35]

Any of the other products presented here might be helpful. Since she is of childbearing age, an appropriate initial recommendation would be riboflavin or magnesium. At present, there are only small studies of all the popular migraine agents, but the studies for riboflavin, magnesium, and coenzyme Q10 are promising. Feverfew is the best-known natural product for migraine prophylaxis; however, the data to support the use of feverfew are contradictory. Feverfew is safe for most patients and if Carol wishes to try it, she should choose a product made from whole leaf prepared in a formulation she can swallow (i.e., she should not chew feverfew leaves.)

Interestingly, there has been a study of a combination product of magnesium, riboflavin, and feverfew that did not appear to offer benefit; however, the study design was poor.[36] If a patient wishes to combine any of the products discussed here, it would be prudent to suggest one agent at a time, so that efficacy of each individual agent can be assessed. If a patient receives benefit from two or more individual agents, a combination product might simplify the regimen.

Self-Assessment

1. Which is TRUE concerning feverfew:
   a. Several large, randomized trials have demonstrated that feverfew reduces migraine frequency by 85%.
   b. Patients should be advised to chew whole-leaf product for best effect.
   c. Feverfew may increase bleeding risk in patients on anticoagulants.
   d. All of the above.

2. Which of the following products is a member of the ragweed family?
   a. Feverfew
   b. Magnesium
   c. Riboflavin
   d. Coenzyme Q10

3. Which is the correct dose of feverfew to recommend in migraine prophylaxis?
   a. 1 to 2 mg/day
   b. 80 to 100 mg/day
   c. 0.2 mcg/day
   d. None of the above

4. Which is TRUE concerning butterbur:
   a. Preliminary data suggest that butterbur may reduce migraine frequency.
   b. Butterbur is typically dosed at 75 mg twice daily.
   c. Butterbur supplements must be pyrrolizidone-free.
   d. All of the above.

5. Which is TRUE concerning riboflavin:
   a. Riboflavin is vitamin B2.
   b. Riboflavin is important in the production of ATP.
   c. Riboflavin 400 mg/day may reduce migraine frequency.
   d. All of the above.

6. Which is TRUE concerning riboflavin:
   a. Riboflavin is very expensive.
   b. Riboflavin deficiency is common.
   c. Riboflavin is inexpensive and well tolerated.
   d. None of the above.

7. **Which is TRUE concerning coenzyme Q10:**
   a. It is also called ubiquinone.
   b. It is important in mitochondrial energy production.
   c. Endogenous levels decrease with aging.
   d. All of the above.

8. **Which is TRUE concerning coenzyme Q10:**
   a. It may increase blood glucose levels.
   b. It is difficult for most patients to tolerate at recommended doses.
   c. A product of good quality is hard to find.
   d. None of the above.

9. **A patient wishes to try a natural product for migraine prophylaxis. She has diarrhea-predominant irritable bowel syndrome and liver disease (hepatitis B and C positive), from years of IV drug abuse. She reports ragweed allergy. Which of the following products is best to recommend?**
   a. Magnesium
   b. Feverfew
   c. Riboflavin
   d. Butterbur

10. **A reasonable trial period for natural products for migraine is 4 to 6 weeks, except for butterbur, which may take longer.**
    a. True
    b. False

Answers: 1-c; 2-a; 3-b; 4-d; 5-d; 6-c; 7-d; 8-d; 9-c; 10-a

REFERENCES

1. Silberstein SD. Migraine. *Lancet.* 2004 Jan 31;363(9406):381–91. Review.
2. de Weerdt CJ, Bootsma HPR, Hendriks H. Herbal medicines in migraine prevention. Randomized double-blind placebo-controlled crossover trial of a feverfew preparation. *Phytomedicine.* 1996;3:225–30.
3. Hwang D, Fischer NH, Jang BC, et al. Inhibition of the expression of inducible cyclooxygenase and proinflammatory cytokines by sesquiterpene lactones in macrophages correlates with the inhibition of MAP kinases. *Biochem Biophys Res Commun.* 1996 Sep 24;226(3):810–8.
4. Heptinstall S, White A, Williamson L, et al. Extracts of feverfew inhibit granule secretion in blood platelets and polymorphonuclear leucocytes. *Lancet.* 1985 May 11;1(8437):1071–4.
5. Barsby RW, Salan U, Knight DW, et al. Feverfew extracts and parthenolide irreversibly inhibit vascular responses of the rabbit aorta. *J Pharm Pharmacol.* 1992;44:737–40.

6. Guzman ML, Rossi RM, Karnischky L, et al. The sesquiterpene lactone parthenolide induces apoptosis of human acute myelogenous leukemia stem and progenitor cells. *Blood.* 2005 Jun 1;105(11):4163–9. Epub 2005 Feb 1.

7. Sweeney CJ, Mehrotra S, Sadaria MR, et al. The sesquiterpene lactone parthenolide in combination with docetaxel reduces metastasis and improves survival in a xenograft model of breast cancer. *Mol Cancer Ther.* 2005 Jun;4(6):1004–12.

8. Murphy JJ, Heptinstall S, Mitchell JR. Randomized, double-blind, placebo-controlled trial of feverfew in migraine prevention. *Lancet.* 1988;2:189–92.

9. Palevitch D, Earon G, Carasso R. Feverfew (*Tanacetum parthenium*) as a prophylactic treatment for migraine—a double-blind, placebo-controlled study. *Phytotherapy Res.* 1997;11:508–11.

10. Johnson ES, Kadam NP, Hylands DM, et al. Efficacy of feverfew as prophylactic treatment of migraine. *Br Med J* (Clin Res Ed). 1985;291:569–73.

11. Vogler BK, Pittler MH, Ernst E. Feverfew as a preventive treatment for migraine: a systematic review. *Cephalalgia.* 1998;18:704–8.

12. Jellin JM, Gregory PJ, Batz F, et al. Natural Medicines Comprehensive Database. Therapeutic Research Faculty. Stockton CA; 2005. Available at www.naturaldatabase.com. Accessed August 6, 2005.

13. Bickel D, Röder T, Bestmann HJ, et al. Identification and characterization of inhibitors of peptido-leukotriene synthesis from *Petasites hybridus*. *Planta Med.* 1994;60:318–22.

14. Thomet OAR, Wiesmann UN, Schapowal A, et al. Role of petasine in the potential anti-inflammatory activity of a plant extract of *Petasites hybridus*. *Biochem Pharmacol.* 2001;61:1041–7.

15. Jellin JM, Gregory PJ, Batz F, et al. Natural Medicines Comprehensive Database. Therapeutic Research Faculty. Stockton CA; 2005. Available at www.naturaldatabase.com. Accessed July 27, 2005.

16. Diener HC, Rahlfs VW, Danesch U. The first placebo-controlled trial of a special butterbur root extract for the prevention of migraine: reanalysis of efficacy criteria. *Eur Neurol.* 2004;51:89–97.

17. Lipton RB, Gobel H, Einhaupl KM, et al. Petasites hybridus root (butterbur) is an effective preventive treatment for migraine. *Neurology.* 2004;63:2240–4.

18. Lee DK, Gray RD, Robb FM, et al. A placebo-controlled evaluation of butterbur and fexofenadine on objective and subjective outcomes in perennial allergic rhinitis. *Clin Exp Allergy.* 2004;34:646–9.

19. Brown DJ. Standardized butterbur extract for migraine treatment: a clinical overview. *HerbalGram.* 2003;58:18–9.

20. Schoenen J, Lenaerts M, Bastings E. High-dose riboflavin as a prophylactic treatment of migraine: results of an open pilot study. *Cephalalgia.* 1994;14:328–9.

21. Boehnke C, Reuter U, Flach U, et al. High-dose riboflavin treatment is efficacious in migraine prophylaxis: an open study in a tertiary care centre. *Eur J Neurol.* 2004;11:475–7.

22. A Report of the Standing Committee on the Scientific Evaluation of Dietary Reference Intakes and Its Panel on Folate, Other B Vitamins, and Choline and Subcommittee on Upper Reference Levels of Nutrients, Food and Nutrition Board, Institute of Medicine. Dietary Reference. National Academy Press. 1998.

23. Rozen TD, Oshinsky ML, Gebeline CA, et al. Open label trial of coenzyme Q10 as a migraine preventive. *Cephalalgia.* 2002;22:137–41.

24. Sandor PS, Di Clemente L, Coppola G, et al. Efficacy of coenzyme Q10 in migraine prophylaxis: a randomized controlled trial. *Neurology.* 2005;64:713–5.

25. Bleske BE, Willis RA, Anthony M, et al. The effect of pravastatin and atorvastatin on coenzyme Q10. *Am Heart J.* 2001;142:E2–7.

26. Engelsen J, Nielsen JD, Winther K. Effect of coenzyme Q10 and Ginkgo biloba on warfarin dosage in stable, long-term warfarin-treated outpatients. A randomised, double-blind, placebo-crossover trial. *Thromb Haemost.* 2002;87:1075–6.

27. Weglicki W, Ouamme G, Tucker K, et al. Potassium, magnesium, and electrolyte imbalance and complications in disease management. *Clin Exp Hypertens.* 2005 Jan;27(1):95–112. Review.

28. Facchinetti F, Sances G, Borella P, et al. Magnesium prophylaxis of menstrual migraine: effects on intracellular magnesium. *Headache.* 1991;31:298–301.

29. Peikert A, Wilimzig C, Kohne-Volland R. Prophylaxis of migraine with oral magnesium: results from a prospective, multi-center, placebo-controlled and double-blind randomized study. *Cephalalgia.* 1996;16:257–63.

30. Taubert K. Magnesium in migraine. Results of a multicenter pilot study. *Fortschr Med.* 1994 Aug 30;112(24):328–30. German.

31. Birrer RB, Shallash AJ, Totten V. Hypermagnesemia-induced fatality following epsom salt gargles. *J Emerg Med.* 2002;22:185–8.

32. McGuire JK, Kulkarni MS, Baden HP. Fatal hypermagnesemia in a child treated with megavitamin/megamineral therapy. *Pediatrics.* 2000;105:E18.

33. Product review: Magnesium. ConsumerLab Web site. Available at www.consumerlab.com. Accessed July 27, 2005.

34. Lake AE. Behavioral and nonpharmacologic treatments of headache. *Med Clin North Am.* 2001 Jul;85(4):1055–75.

35. Savi L, Rainero I, Valfre W, et al. Food and headache attacks. A comparison of patients with migraine and tension-type headache. *Panminerva Med.* 2002 Mar;44(1):27–31.

36. Maizels M, Blumenfeld A, Burchette R. A combination of riboflavin, magnesium, and feverfew for migraine prophylaxis: a randomized trial. *Headache.* 2004;44:885–90.

CHAPTER 4

Osteoarthritis

Karen Shapiro

Osteoarthritis (OA) is the most common type of arthritis and the most common cause of chronic pain and disability in older adults.[1] Osteoarthritis is also referred to as degenerative joint disease. It occurs most commonly in the fingers, hips, knees, and spine. Symptoms include joint aching and soreness, pain, and bony knots in the finger joints.

This is the leading medical condition for which people use alternative therapies.[2] Acetaminophen is the drug of choice in initial cases of osteoarthritis; however, the analgesia provided by acetaminophen is often insufficient.[3] Nonsteroidal anti-inflammatory drugs (NSAIDs) such as ibuprofen are commonly used; however, these are not without clinical concern. This is especially true when NSAIDs are taken chronically in high doses, as is commonly done with this condition. Patients often turn to alternative therapies due to incomplete symptom relief or after having bothersome side effects with conventional treatments. Many patients use conventional and alternative therapies concurrently.

KEY POINTS

- Substantial data support the use of glucosamine sulfate for symptom improvement and disease-modifying effects in osteoarthritis.
- Chondroitin sulfate has considerable data supporting use in osteoarthritis.
- S-Adenosylmethionine (SAMe) has data supporting use in both osteoarthritis and depression.
- SAMe should not be used concurrently with drugs that increase serotonin transmission.
- Capsaicin reduces pain caused by hand and knee osteoarthritis.
- Capsaicin is nonsystemic and safe; however, care must be used in application.
- Fish oil (omega-3 fatty acids) is promising for osteoarthritis, but data are insufficient at this time to recommend supplements for this purpose.
- Avocado and soybean unsaponifiables (ASU) are promising for osteoarthritis.
- Glucosamine, chondroitin, S-adenosylmethionine, capsaicin, avocado and soybean oils, and fish oil do not offer fast symptom relief and should be used for a reasonable trial period.
- Doses for natural products, similar to prescription drugs, should match available evidence.

Product	Dosage	Effect	Safety Concerns
Glucosamine	1500 mg/day or divided; trial period of at least 4 weeks	Effective for symptom reduction; evidence for disease modification	• Patients with diabetes should monitor blood glucose
Chondroitin	1200 mg/day, divided twice or three times daily; trial period of at least 4 weeks	Effective for symptom reduction	• May have weak anticoagulant activity • Avoid if history of prostate cancer
S-Adenosylmethionine (SAMe)	400 to 600 mg/day, taken in divided doses; trial period of at least 4 weeks	Effective for symptom reduction	• Significant variations in quality, expensive • Avoid with serotonergic drugs • Avoid in bipolar disorder
Capsaicin	0.025% or 0.075%, applied four times a day; trial period of 2 to 4 weeks	Effective for pain relief for hand and knee osteoarthritis; useful for musculoskeletal and neuropathic conditions	• Burning sensation when beginning therapy • Avoid contact with mucous membranes
Avocado and soybean oils (ASU)	300 mg once daily	Promising data for symptom relief and improvements in joint space narrowing	• None
Fish oil (Omega-3 fatty acids)	Increase consumption in food; supplemental doses start at 500 mg eicosapentaenoic acid (EPA) and docosahexaenoic acid (DHA) daily	May be helpful for anti-inflammatory effects; may decrease chondrocyte degradation; preliminary data	• Can increase bleeding risk with high doses • Can cause fishy taste at high doses

GLUCOSAMINE AND CHONDROITIN

Glucosamine is the most well known of natural products used for osteoarthritis and has a long history of use in veterinary medicine, where it is commonly used for race horses who suffer osteoarthritic knee damage. The supplement became popular for human use in 1997 after Jason Theodosakis, MD, advocated the benefits of glucosamine in the bestselling book *The Arthritis Cure*.

Glucosamine is produced naturally in the body, beginning with the addition of an amino group to glucose (and thus the name, glucose-amine). Glucosamine is present in almost all human tissues and is concentrated in cartilage and other connective tissues. Glucosamine sold commercially is isolated from chitin, a substance found in crustacean shells, including crab and oyster, or is made synthetically.

Glucosamine serves as the starting substrate in the synthesis of glycosaminoglycans and proteoglycans—vital components of cartilage that give joints their "shock-absorber"-like effect. The glycosaminoglycans in articular

PATIENT CASE

ELLEN

Ellen is a 57-year-old, moderately overweight white female who complains of knee pain, especially after going on a long walk or carrying something heavy. Her knees hurt worse at the end of the day. Her knees first began to bother her about 2 years ago. She relates limitation of motion and states that sometimes she can't squat down to pick up something heavy because her knees feel stiff.

Ellen complains of joint swelling, most notably in the right knee for about the past 2 months. The right knee was injured about 2 months ago when she fell on it. This knee has been very warm recently, and there was some joint redness during the past week. Her right knee also makes a noise when she bends it. The noise is described as "grating." Ellen denies back pain, neck pain, sprains, or deformities. Plain film radiograph of both knees shows joint narrowing.

Ellen had used acetaminophen for the first year with acceptable pain relief. She reports that it no longer helps. She finds some relief from ibuprofen 600 mg, taken three times daily, but now she is worried about taking these pills because of recent news stories. She has decided to try something natural.

Her other medications include fluoxetine 20 mg/day, amitriptyline 50 mg nightly, and sumatriptan 50 mg as needed for migraine.

cartilage include chondroitin sulfate and other compounds. In osteoarthritis, the loss of the glycosaminoglycan layer results in articular cartilage erosion with subsequent pain and functional impairment. Glucosamine supplements serve as a substrate for the synthesis of glycosaminoglycans and may stimulate the synthesis of glycosaminoglycans and proteoglycans by chondrocytes.[4,5] Glucosamine may also inhibit cartilage breakdown.

The majority of clinical trials demonstrate positive results.[6,7] A recent randomized trial of 414 postmenopausal women with knee osteoarthritis demonstrated disease-modifying effects. Subjects received 1500 mg/day of glucosamine sulfate or placebo and were followed for 3 years. In the glucosamine group, 6.9% had significant joint space narrowing of more than 0.5 mm compared with 20.6% of women in the placebo group (relative risk for glucosamine group: 0.33). Subjects in the glucosamine group had a 14.1% improvement in clinical symptoms after 3 years. During the same period, the placebo group's symptom scores worsened by 5.4%.[8] Disease-modifying benefit is important since osteoarthritis is a progressive disease and no standard agent provides benefit beyond pain reduction.

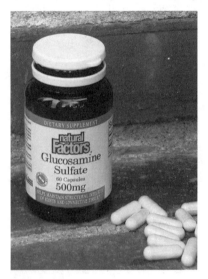

Glucosamine

Results reported from the NIH-sponsored Glucosamine–Chondroitin Arthritis Intervention Trial in 2005 were positive for patients with moderate-to-severe knee osteoarthritis pain.[9] Almost 1600 patients at 16 U.S. academic sites were randomized to receive glucosamine hydrochloride, chondroitin sulfate, both supplements, celecoxib, or placebo. In the glucosamine–chondroitin combination group, 79.2% of subjects experienced pain relief. For patients taking celecoxib, 69.4% experienced pain relief. The glucosamine-alone group and the chondroitin-alone group had a 65.7% and 61.4% pain reduction, respectively. The placebo group reported a 54.3% reduction in pain. In patients with mild knee osteoarthritis pain, the glucosamine–chondroitin combination was not significantly more effective than placebo. It is possible that small differences in the patients with mild pain, if present, were not detected.

Chondroitin sulfate is a heparin-like, complex carbohydrate. In vivo and in vitro studies demonstrate that chondroitin sulfate stimulates chondrocyte synthesis of collagen, proteoglycans, and other cartilage components.[10] Chondroitin also improves articular water retention, improves elasticity, and inhibits enzymatic degradation of cartilage. Some of the effects are attributable to anti-inflammatory properties.[11] Chondroitin sulfate, as a single agent, has been shown to decrease pain and improve knee function.[12-14] Commercially available chondroitin is derived mostly from shark and cow cartilage.[2]

The typical daily dose of glucosamine used in clinical trials is 1500 mg, taken once or in divided doses. The typical dose of chondroitin sulfate is 1200 mg/day, taken in two or three divided doses. The two agents have different mechanisms of action, and it is reasonable (although not necessary) to administer them in combination products for additive benefit.

SAFETY CONSIDERATIONS/DRUG INTERACTIONS

Long-term treatment with glucosamine is well tolerated, with a safety profile superior to NSAIDs. Most major studies of glucosamine used the sulfate salt and other forms of glucosamine may not be as effective. A trial of at least 4 weeks is necessary to assess benefit. NSAIDs provide a faster reduction in pain, but glucosamine's effects last longer and symptoms do not recur soon after discontinuation, as they do with NSAIDs. NSAIDs may need to be continued, at least initially, due to the slow onset of action. Patients with diabetes should be counseled to monitor blood glucose levels carefully due to case reports of increased glucose levels in patients taking glucosamine. However, significant changes in blood glucose levels have not been found in clinical trials.

Chondroitin may have weak anticoagulant activity and should be used cautiously in patients taking warfarin.[15] Men with prostate cancer should avoid taking chondroitin.[16] A component in chondroitin called versican is overexpressed in prostate cancer and may contribute to disease progression.[17] Like glucosamine, chondroitin does not have immediate effects and a reasonable trial period of 4 weeks should be recommended.

S-ADENOSYLMETHIONINE (SAME)

SAMe (pronounced Sam-ee) is present in all living cells and acts as a methyl donor in reactions that are essential in the production of a wide range of compounds, including dopamine, serotonin, and cartilage proteoglycans. It is thought that when endogenous levels of SAMe are low, supplements increase production of these compounds. SAMe is used as a prescription drug in Europe for osteoarthritis and depression. SAMe has been studied for osteoarthritis of the knee, hip, spine, and hand. It is better tolerated and as effective as NSAIDs for improving function and for decreasing pain.[18-22] A recent study compared SAMe at a higher-than-typical dosage (1200 mg/day) to celecoxib (200 mg/day) in a double-blind, crossover trial and found similar efficacy for pain reduction.[23]

Safety Considerations/Drug Interactions

In this trial[23] as in others, SAMe had a slower onset of action for pain relief than the NSAID comparator. A 4-week trial period is reasonable. Like glucosamine and chondroitin, the slower onset of action may require an initial period of "bridge" therapy with NSAIDs. Other patients, with more advanced disease, will need to continue NSAID use, but they may be able to reduce the dose.

SAMe is expensive to produce. The higher the cost of a supplement to produce, the higher the risk of poor quality. SAMe is no exception and there is significant variance in both quality and quantity.[24] A stabilizing compound is added to SAMe that can make calculating dosage difficult. Patients should use the ConsumerLab Web site or another reputable source to pick an appropriate product. SAMe products are expensive and a day's supply ranges from $1.00 to $4.00. A typical daily dose of SAMe for osteoarthritis is 400 to 600 mg/day, divided in two or three doses.

SAMe can elevate neurotransmitters and enhance serotonin transmission. It should not be used with other drugs that do the same, including many antidepressants, since the combination could result in serotonin syndrome. Patients with bipolar disorder should not use SAMe due to the possibility that the compound could trigger a manic episode.[25]

Capsaicin

Capsaicin creams and ointments are popular with patients for hand and knee osteoarthritis. Capsaicin is a substance from hot peppers that causes tissues to release substance P, a pain transmitter. When capsaicin is applied topically, the substance P becomes depleted and pain transmission to the brain is decreased. Capsaicin reduces but does not eliminate pain due to musculoskeletal and neuropathic conditions.[26,27] This makes it useful for adjunctive therapy in patients with moderate-to-severe pain or as a possible single agent in mild pain.

Patients who are wary of taking systemically acting agents are often willing to use topical agents.[28] Capsaicin comes in two strengths, 0.025% (Zostrix® and generics)

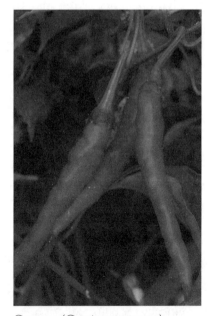

Cayenne (Capsicum annuum)

and 0.075% (Zostrix® HP and generics). Patients should begin with the lower strength formulation, applied four times daily, and then switch, if needed, to the higher formulation once the burning sensation is diminished. It takes some time to deplete substance P and consequently to notice significant pain relief. Patients will need to give capsaicin a 2- to 4-week trial period.

Safety Considerations/Drug Interactions

Capsaicin, when applied topically, causes a burning sensation that diminishes over time. If the hands are being treated, instruct patients to leave the medication on for 30 minutes and then wash hands well with soap and water. Otherwise, wear gloves for the application or wash hands well afterward. Topical capsaicin should not be covered with bandages or heat wraps. Touching capsaicin-covered fingers to eyes, lips, and genital areas will cause considerable pain. Capsaicin, used topically, has no significant health concerns and no known drug interactions.

Avocado and Soybean Oils

Avocado and soybean unsaponifiables (ASU) are promising for osteoarthritis symptom relief and improvement in joint space narrowing. This product is thought to induce synthesis of the cartilage matrix and prevent cartilage degradation.[29,30] A majority of trial data, including trials with high methodological quality, show benefit for hip and knee osteoarthritis.[31-34] Another benefit shown in these studies is a significant decrease in NSAID use.

The dose is one 300-mg tablet daily. Avocado and soybean unsaponifiables is marketed alone and in combination with glucosamine and chondroitin.

Safety Considerations/Drug Interactions

This product is well tolerated with no known drug interactions.

Fish Oil (Omega-3 Fatty Acids)

Omega-3 fatty acids are discussed in some detail in Chapters 10 and 11. A primary mechanism of action for omega-3 fatty acids in cardiovascular disease and other disease states, including rheumatoid arthritis, involves its anti-inflammatory properties. In osteoarthritis, an inflammatory component is not typically present initially but may develop over time. This is the reason why acetaminophen may be sufficient, when used at appropriate doses, for early osteoarthritis but can later become insufficient for pain control. At this time, patients are typically switched to NSAIDs. Preliminary data indicate that omega-3 fatty acid supplementation may decrease inflammation in osteoarthritis and decrease chondrocyte degradation.[35,36] Further studies are required to demonstrate benefit.

In the meantime, it is reasonable to increase consumption of omega-3 fatty acids for most people for general health benefits and possibly for benefit in osteoarthritis. It is believed that the typical western-style diet is too low in these types of fats and much too high in unhealthy fats. A patient who wishes to consume more fatty fish, cooked in a healthful manner, should be encouraged. Supplementation at the present time for osteoarthritis is premature. Typical supplemental daily doses used for other indications are 500 mg or higher.

SAFETY CONSIDERATIONS/DRUG INTERACTIONS

Omega-3 fatty acids are a natural part of the human diet and are free of major side effects. The U.S. Food and Drug Administration has ruled that intake of up to 3 g/day of fish oil is generally recognized as safe. An unpleasant fishy aftertaste may be experienced with high doses of supplements. There is a potential for increased risk of bleeding when combined with drugs like warfarin or aspirin, but the evidence for bleeding has not been documented with daily doses less than 3 g.

Concerns for exposure to environmental contaminants (e.g., methylmercury, polychlorinated biphenyls, and dioxins) associated with increased fish consumption should be considered. All persons should attempt to limit consumption of fish that have high levels of methylmercury. Children and pregnant and lactating women should avoid these fish, which include swordfish, shark, king mackerel, and tilefish.

PATIENT DISCUSSION

Ellen reports that acetaminophen had helped in the past but is no longer adequate. Her condition is notable for inflammation, as is evident by the redness, swelling, and warmth. Ellen is using an NSAID for pain relief. The chronic use of these agents is not without clinical concern, including gastrointestinal and renal toxicities. There is another pressing issue in Ellen's case. Neither acetaminophen nor NSAIDs will improve or halt the disease process. Use of glucosamine can result in joint improvement. A combination product with chondroitin, which also has data to support efficacy, would be reasonable. A trial of SAMe would not be appropriate for this patient. She is taking several medications currently that raise serotonin and norepinephrine levels.

If she wishes to increase her consumption of fish oils, this is reasonable, particularly as an inflammatory component is present. Fish oils can stabilize cardiac conductivity and she is taking amitriptyline, a drug that can cause QT prolongation. However, the dose of the medication is low, she is not taking additive QT-prolonging agents nor is she known to have preexisting cardiovascular disease. This is interesting to note but not reason enough to take fish oil supplements. Avocado and soybean unsaponifiables is another option that is safe and can be given a trial if the patient wishes to try this agent.

She should be counseled to use only one product (or two, if a glucosamine–chondroitin combination product is used) at a time and to give the trial a reasonable length of time. Otherwise, there is no way to tell which ingredient/s are causing benefit or possibly harm. All agents discussed here should be used routinely (not as needed) and in recommended doses.

Ellen is overweight, and she should be assisted with a weight-loss program to reduce stress on her knees and other weight-bearing joints. Physical activity should be encouraged, including range-of-motion exercises, low-impact aerobic exercise, such as walking, and muscular strengthening, in order to help prevent worsening of her condition.[37]

SELF-ASSESSMENT

1. A patient has osteoarthritis with an inflammatory component and wishes to try a natural product. Which of the following natural products has anti-inflammatory effects?
   a. Capsaicin
   b. Fish oils
   c. Chondroitin
   d. B and C

2. Which of the following products raises serotonin levels?
   a. Glucosamine
   b. SAMe
   c. Avocado and soybean oils
   d. Fish oils

3. Which of the following products should be avoided in patients with a history of prostate cancer?
   a. Chondroitin
   b. SAMe
   c. Avocado and soybean oils
   d. Fish oils

4. Which of the following products has weak anticoagulant activity?
   a. Chondroitin
   b. SAMe
   c. Glucosamine
   d. Fish oils

5. Capsaicin depletes a transmitter that sends pain signals to the CNS. This transmitter is called:
   a. Substance C
   b. Substance P
   c. Enkephalin
   d. Serotonin

6. Many natural products require a lag time for benefit. However, if a patient is using nonprescription or prescription NSAID therapy, the NSAID should be discontinued when the natural product is started. This is necessary to avoid the possibility of harmful interactions.
   a. True
   b. False

7. Chondroitin, but not glucosamine, can be taken on an as-needed basis.
   a. True
   b. False

8. A 53-year-old Asian female presents with complaints of hand osteoarthritis. She went to her doctor who recommended the prescription NSAID indomethacin. She tried indomethacin and states that the pain was improved. However, she stopped taking it because she felt disoriented and got an upset stomach. She is wary of taking anything else systemically. An appropriate recommendation for this patient could include:
   a. Capsaicin 0.025%, applied topically, with a possible increase to the higher strength once initial pain upon application is diminished
   b. Capsaicin 0.075%, with a possible decrease to the lower strength once pain is resolved
   c. Avocado oil, applied topically
   d. None of the above

9. All patients using glucosamine need regular glucose monitoring.
   a. True
   b. False

10. Which of the following products has a high risk of poor product quality and is expensive?
    a. Avocado and soybean oils
    b. SAMe
    c. Chondroitin
    d. Glucosamine

Answers: 1-d; 2-b; 3-a; 4-a; 5-b; 6-b; 7-b; 8-a; 9-b; 10-b

REFERENCES

1. Loeser RF Jr. Aging cartilage and osteoarthritis—what's the link? *Sci Aging Knowledge Environ*. 2004 Jul 21;2004(29):pe31.

2. Morelli V, Naquin C, Weaver V. Alternative therapies for traditional disease states: osteoarthritis. *Am Fam Phys*. 2003;67(2):339–44.

3. Case JP, Baliunas AJ, Block JA. Lack of efficacy of acetaminophen in treating symptomatic knee osteoarthritris: a randomized, double-blind, placbo-controlled comparison trial with diclofenac sodium. *Arch Intern Med*. 2003;163:169–78.

4. Dodge GR, Regatte RR, Hall JO, et al. The fate of oral glucosamine traced by C-labeling in the dog. 65th American College of Rheumatology Meeting, San Francisco, CA; November 2001. Abstract.

5. Bassleer C, Rovati L, Franchimont P. Stimulation of proteoglycan production by glucosamine sulfate in chondrocytes isolated from human osteoarthritic articular cartilage in vitro. *Osteoarthritis Cartilage*. 1998 Nov;6(6):427–34.

6. Richy F, Bruyere O, Ethgen O, et al. Structural and symptomatic efficacy of glucosamine and chondroitin in knee osteoarthritis: a comprehensive meta-analysis. *Arch Intern Med*. 2003;163:1514–22.

7. Braham R, Dawson B, Goodman C. The effect of glucosamine supplementation on people experiencing regular knee pain. *Br J Sports Med*. 2003;37:45–9.

8. Bruyere O, Pavelka K, Rovati LC, et al. Glucosamine sulfate reduces osteoarthritis progression in postmenopausal women with knee osteoarthritis: evidence from two 3-year studies. *Menopause*. 2004 Mar-Apr;11(2):138–43.

9. ACR 2005 Meeting Abstract on GAIT trial. Available at www.rheumatology.org/research. Accessed December 10, 2005.

10. Kelly GS. The role of glucosamine sulfate and chondroitin sulfates in the treatment of degenerative joint disease. *Altern Med Rev*. 1998;3:27–39.

11. Ronca F, Palmieri L, Panicucci P, et al. Anti-inflammatory activity of chondroitin sulfate. *Osteoarthritis Cartilage*. 1998;6 (Suppl A):14–21.

12. Uebelhart D, Malaise M, Marcolongo R, et al. Intermittent treatment of knee osteoarthritis with oral chondroitin sulfate: a one-year, randomized, double-blind, multicenter study versus placebo. *Osteoarthritis Cartilage*. 2004 Apr;12(4):269–76.

13. Uebelhart D, Thonar EJ, Delmas PD, et al. Effects of oral chondroitin sulfate on the progression of knee osteoarthritis: a pilot study. *Osteoarthritis Cartilage*. 1998 May;6 (Suppl A):39–46.

14. Scroggie DA, Albright A, Harris MD. The effect of glucosamine–chondroitin supplementation on glycosylated hemoglobin levels in patients with type 2 diabetes mellitus: a placebo-controlled, double-blinded, randomized clinical trial. *Arch Intern Med*. 2003;163:1587–90.

15. Rozenfeld V, Crain JL, Callahan AK. Possible augmentation of warfarin effect by glucosamine–chondroitin. *Am J Health Syst Pharm.* 2004;61:306–7.

16. Ricciardelli C, Quinn DT, Raymond WA, et al. Elevated levels of peritumoral chondroitin sulfate are predictive of poor prognosis in patients treated by radical prostatectomy for early-stage prostate cancer. *Cancer Res.* 1999;59:2324–8.

17. Cross NA, Chandrasekharan S, Jokonya N, et al. The expression and regulation of ADAMTS-1, -4, -5, -9, and -15, and TIMP-3 by TGFbeta1 in prostate cells: relevance to the accumulation of versican. *Prostate.* 2005 May 15;63(3):269–75.

18. Bradley JD, Flusser D, Katz BP, et al. A randomized, double blind, placebo controlled trial of intravenous loading with S-adenosylmethionine (SAM) followed by oral SAM therapy in patients with knee osteoarthritis. *J Rheumatol.* 1994;21:905–11.

19. Caruso I, Petrogrande V. Italian double-blind multicenter study comparing S-adenosylmethionine, naproxen and placebo in the treatment of degenerative joint disease. *Am J Med.* 1987;83(5A):66–71.

20. Glorioso S, Todesco S, Mazzi A, et al. Double-blind multicentre study of the activity of S-adenosylmethionine in hip and knee osteoarthritis. *Int J Clin Pharmacol Res.* 1985;5:39–49.

21. Vetter G. Double-blind comparative clinical trial with S-adenosylmethionine and indomethacin in the treatment of osteoarthritis. *Am J Med.* 1987;83(5A):78–80.

22. Muller-Fassbender H. Double-blind clinical trial of S-adenosylmethionine versus ibuprofen in the treatment of osteoarthritis. *Am J Med.* 1987;83(5A):81–3.

23. Najm WI, Reinsch S, Hoehler F, et al. S-Adenosyl methionine (SAMe) versus celecoxib for the treatment of osteoarthritis symptoms: a double-blind crossover trial. [ISRCTN36233495]. *BMC Musculoskelet Disord.* 2004 Feb 26;5(1):6.

24. SAMe study results. Available at www.consumerlab.com. Accessed December 7, 2004.

25. Lipinski JF, Cohen BM, Frankenburg F, et al. Open trial of S-adenosylmethionine for treatment of depression. *Am J Psychiatry.* 1984 Mar;141(3):448–50.

26. Mason L, Moore RA, Derry S, et al. Systematic review of topical capsaicin for the treatment of chronic pain. *BMJ.* 2004 Apr 24;328(7446):991. Epub Mar 19, 2004. Review.

27. Tramer MR. It's not just about rubbing—topical capsaicin and topical salicylates may be useful as adjuvants to conventional pain treatment. *BMJ.* 2004 Apr 24;328(7446):998.

28. Fraenkel L, Bogardus ST Jr, Concato J, et al. Treatment options in knee osteoarthritis: the patient's perspective. *Arch Intern Med.* 2004 Jun 28;164(12):1299–304.

29. Reginster J, Gillot V, Bruyere O, et al. Evidence of nutriceutical effectiveness in the treatment of osteoarthritis. *Curr Rheumatol Rep.* 2000;2:472–7.

30. Henrotin YE, Sanchez C, Deberg MA, et al. Avocado/soybean unsaponifiables increase aggrecan synthesis and reduce catabolic and proinflammatory mediator production by human osteoarthritic chondrocytes. *J Rheumatol.* 2003;30:1825–34.

31. Blotman F, Maheu E, Wulwik A, et al. Efficacy and safety of avocado/soybean unsaponifiables in the treatment of symptomatic osteoarthritis of the knee and hip. A prospective, multicenter, three-month, randomized, double-blind, placebo-controlled trial. *Rev Rhum Engl Ed.* 1997 Dec;64(12):825–34.

32. Lequesne M, Maheu E, Cadet C, et al. Structural effect of avocado/soybean unsaponifiables on joint space loss in osteoarthritis of the hip. *Arthritis Rheum.* 2002 Feb;47(1):50–8.

33. Ernst E. Avocado–soybean unsaponifiables (ASU) for osteoarthritis—a systematic review. *Clin Rheumatol.* 2003 Oct;22(4-5):285–8. Review.

34. Maheu E, Mazieres B, Valat JP, et al. Symptomatic efficacy of avocado/soybean unsaponifiables in the treatment of osteoarthritis of the knee and hip: a prospective randomized, double-blind, placebo-controlled, multicenter trial with a six-month treatment period and a two-month followup demonstrating a persistent effect. *Arthritis Rheum.* 1998;41:81–91.

35. Curtis CL, Rees SG, Cramp J, et al. Effects of n-3 fatty acids on cartilage metabolism. *Proc Nutr Soc.* 2002 Aug;61(3):381–9.

36. Curtis CL, Rees SG, Little CB, et al. Pathologic indicators of degradation and inflammation in human osteoarthritic cartilage are abrogated by exposure to n-3 fatty acids. *Arthritis Rheum.* 2002 Jun;46(6):1544–53.

37. Roddy E, Zhang W, Doherty M. Aerobic walking or strengthening exercise for osteoarthritis of the knee? A systematic review. *Ann Rheum Dis.* 2005;64:544–8.

Chapter 5
Erectile Dysfunction

Karen Shapiro

Erectile dysfunction (ED) is the inability to achieve and sustain an erection suitable for sexual intercourse; 10% of adult males suffer from long-term erectile dysfunction. Many other men have occasional problems with achieving or maintaining an erection, which can be due to many reasons, including stress, exhaustion, or drinking too much alcohol. If the problem persists, the patient should visit a physician to get a physical examination to diagnose erectile dysfunction, to determine any underlying causes, and to identify appropriate treatment. Different chronic illnesses and certain medications can cause erectile dysfunction, including many antidepressants and antihypertensives.

Erectile dysfunction is due to any one or a combination of these factors: damage to the nerves that innervate the penis, decreased blood circulation to the penis, and/or lack of stimulus from the brain.

During sexual stimulation, nitric oxide (NO) is released. Nitric oxide stimulates cyclic guanosine monophosphate (cGMP). This is the key to getting and

KEY POINTS

- l-Arginine increases nitric oxide levels.
- Higher nitric oxide levels cause increased blood flow to the penis and clitoris, via stimulation of cGMP.
- l-Arginine taken alone for erectile dysfunction may require doses of 5 g/day.
- Lower doses of l-arginine may be effective if taken in combination with pycnogenol.
- l-Arginine is well tolerated.
- DHEA might be appropriate for patients with low DHEA-S levels; at present there is no evidence that it helps persons with normal levels.
- DHEA can cause androgenic and estrogenic side effects.
- DHEA may contribute to hormone-sensitive conditions, including cancers.
- Panax ginseng has been used to increase sexual performance for many years; data to support this use are preliminary.
- There is no evidence to support the use of the natural product yohimbe; there is only evidence to support the use of the prescription drug yohimbine.
- Yohimbine (and possibly yohimbe) can cause many undesirable side effects, including hypertension, tachycardia, anxiety, and insomnia.

sustaining an erection. The cGMP causes an increase in blood flow to the penis (or clitoris), resulting in an erection in men (and clitoral enlargement in women). Sildenafil, a phosphodiesterase type 5 (PDE5) inhibitor, blocks the enzyme (PDE5) that degrades cGMP in the corpus cavernosum.

Phosphodiesterase inhibitors are very effective for erectile dysfunction in men but are not without some clinical concern. They cannot be used with certain other medications, including nitrates, and can (rarely) cause serious problems, including loss of vision.

L-ARGININE

Arginine, or arginine, is an amino acid that is required for the synthesis of nitric oxide. It was hypothesized that an increase in nitric oxide might cause smooth muscle relaxation and increased blood flow. While sildenafil blocks the degradation of cGMP, l-arginine increases nitric oxide synthesis, resulting in higher concentrations of activated cGMP.[1-3]

Product	Dosage	Effect	Safety Concerns
l-Arginine	5 g/day; lower doses not effective	May be beneficial; preliminary evidence; strongest data of available products	• Hypotensive effects when combined with blood pressure lowering drugs • Higher doses have more risk of hypotensive effects • Do not use after a heart attack
DHEA	50 mg/day	Preliminary evidence; may benefit men with low DHEA	• Steroid-like side effects (hirsutism, low HDL, acne), more likely with higher doses • Mania, hepatic dysfunction possible • May increase concentrations of CYP 450 3A4 substrates
Panax ginseng	900 mg three times daily	Preliminary evidence	• Side effects uncommon but could include insomnia, headache, palpitations, tachycardia
Yohimbe	15 to 30 mg/day of yohimbine	No evidence for yohimbe; several studies for purified yohimbine	• Potentially serious side effects; not appropriate for self-treatment

Preliminary data suggest that this mechanism may offer some benefit, although modest compared to the phosphodiesterse inhibitors. In a randomized study of 50 males with organic erectile dysfunction, 31% of patients taking l-arginine 5 g/day reported subjective improvement of sexual function over a period of 6 weeks.[4] Another small randomized study using a much lower dose of 1.5 g/day found no significant benefit compared to placebo.[5] Based on what is known, men who wish to try l-arginine alone should use a dose of 5 g/day.

Preliminary evidence suggests that combining a lower dose of l-arginine (1.7 g/day) with pycnogenol might increase the effectiveness of l-arginine.[6] Pycnogenol is an extract of the bark from the French maritime pine tree. Constituents of this extract are thought to stimulate nitric oxide synthase, which theoretically would increase conversion of l-arginine to nitric oxide.[7]

l-Arginine may offer benefit to women with decreased libido. In a double-blind study of 77 women receiving the combination product ArginMax® (2,500 mg of l-arginine, Panax ginseng, Ginkgo biloba extract, damiana, plus numerous vitamins and minerals) or placebo found that subjects in the treatment group had statistically greater improvement in sexual desire and a higher frequency of orgasms (47% versus 30%).

Safety Considerations/Drug Interactions

l-Arginine is usually well tolerated. It has been safely used in studies lasting up to 6 months. No significant side effects were seen in either treatment or placebo groups. Due to l-arginine's vasodilatory effects, it has the potential

PATIENT CASE

Eric

Eric is a 57-year-old male with atherosclerosis, hypertension, and diabetes. His cholesterol, blood pressure, and blood glucose are well controlled. He has a history of not being able to consistently develop and maintain an erection for the past 2 years. Eric uses Viagra® (sildenafil) for this problem and reports that it works well. His other medications include atenolol, candesartan, glyburide, metformin, pravastatin, and aspirin.

Recently, Eric returned to the office to report that his insurance has limited the Viagra to one pill weekly. He states he cannot afford to purchase many additional pills at a cost of $9.00 each. He is asking if there are any natural forms of Viagra, or anything else that will work well and does not cost as much. Eric may have a combination of neurogenic (nerve) damage and vascular problems (resulting in decreased blood flow), based on his medical conditions.

to lower blood pressure and potentially worsen hypotension. Patients with hypotensive episodes should avoid l-arginine.

Due to l-arginine's vasodilatory effects, it might have additive blood pressure lowering effects when combined with antihypertensives, nitrates, and phosphodiesterase inhibitors such as sildenafil. These combinations should be avoided.

DHEA

It is surprising that DHEA (dehydroepiandrosterone) is considered a dietary supplement. DHEA is a hormone naturally produced by the adrenal gland, testes, and liver. It is converted to androstenedione, which is a major precursor to androgens such as testosterone and estrogens.[8]

DHEA (dehydroepiandrosterone)

Since DHEA is a precursor to androgens, there is interest in using it for erectile dysfunction. But when DHEA is taken as a supplement, it gets converted to androgens and estrogens in a gender-specific manner. When women take it, they tend to have increased levels of androgens, but not a significant increase in estrogens. For men, it is the opposite. Men tend to have more of an increase in estrogen levels and not a significant increase in androgens.[9,11,12] In some people, there is speculation that low levels of DHEA and the sulfate ester, DHEA-S, contribute to erectile dysfunction.[13]

Two randomized clinical trials suggest that DHEA might help some patients with organic erectile dysfunction who have lower DHEA levels. Otherwise healthy patients and patients with hypertension who took DHEA 50 mg/day were better able to achieve full erections satisfactory for sexual performance compared to placebo.[14,15] But DHEA did not seem to benefit patients with diabetes who had erectile dysfunction. It is not known if DHEA is beneficial for patients without low DHEA levels. Interestingly, in both of these studies, serum testosterone levels did not increase significantly following DHEA administration.

Doses of 50 mg/day have been used in trials for erectile dysfunction. It is not known if lower or higher doses would be more effective.

Safety Considerations/Drug Interactions

DHEA has been safely used for up to 6 months in clinical trials. Side effects are generally mild at relatively low doses of 50 mg/day. Higher doses are more likely to cause side effects such as insomnia, hepatic dysfunction, insulin resistance, hair loss, abdominal pain, and hypertension. In women, DHEA has been linked with hirsutism, acne, and changes in menstrual pattern.[16]

Mania has also been reported in three cases, including in people with no history of psychiatric disease and in patients with a history of mania and bipolar disorder. Mania has occurred when doses of 50 to 300 mg/day have been used for 2 to 6 months.[17-19]

Preliminary research suggests that DHEA might increase the growth of cancerous cells.[20,21] But other research suggests the opposite, that DHEA might protect against cancer.[22] Since DHEA is an estrogen precursor, it should be avoided in patients with hormone-sensitive conditions such as breast cancer, uterine cancer, ovarian cancer, endometriosis, polycystic ovary syndrome, or uterine fibroids.

DHEA might also worsen diabetes by increasing insulin resistance. DHEA might also worsen atherosclerosis by lowering protective HDL cholesterol and possibly increasing macrophage foam cells.[7] DHEA also seems to inhibit cytochrome P450 3A4. So far this interaction has only been documented with triazolam. DHEA 200 mg/day significantly increases triazolam levels.[23] Theoretically, DHEA might increase levels of other substrates of CYP 3A4. It is not known to what extent lower doses of DHEA might inhibit CYP 3A4.

Panax Ginseng

Panax ginseng or Asian ginseng has a long history of use as a general tonic for increasing vitality, including sexual performance.

One small, randomized trial suggests that taking a Panax ginseng extract 900 mg three times daily subjectively improves erection in men with erectile dysfunction. Panax ginseng also seems to improve objective measures of penile rigidity.[24] This trial used 900 mg three times daily. There is not enough evidence to determine the optimal dose.

Panax ginseng

Safety Considerations/Drug Interactions

Panax ginseng is usually well tolerated. Most side effects do not occur any more frequently than with placebo. One of the most commonly reported side effects among patients who take Panax ginseng is insomnia. Other side effects that have been reported include tachycardia, mastalgia, vaginal bleeding, decreased appetite, headache, palpitations, vertigo, and itchy skin.[7]

Panax ginseng might also affect cardiac function. There is some evidence that Panax ginseng can prolong the QT interval and might also modestly decrease blood pressure. However, the long-term effects of ginseng on cardiac function are unknown. Panax ginseng should be used cautiously or avoided in patients with cardiovascular conditions.

Panax ginseng seems to lower blood glucose levels. Theoretically, combining ginseng with insulin or hypoglycemic drugs (e.g., glyburide) might increase the risk of hypoglycemia.

Panax ginseng also seems to reduce the effectiveness of warfarin (Coumadin®).[8] Since ginseng seems to stimulate immune function, there is also concern that it might decrease the effectiveness of immunosuppressants such as cyclosporine. However, this interaction has not been documented. Ginseng might also interfere with monoamine oxidase inhibitors. There is a report of insomnia, headache, tremor, and hypomania in a patient who combined an unspecified ginseng species with phenelzine (Nardil®).[8]

Yohimbe

The bark of the evergreen yohimbe tree contains the active constituent yohimbine, which is a prescription drug. The bark contains approximately 6% yohimbine.[25] Surprisingly, this product is still on the market as a dietary supplement despite containing a constituent that is a prescription drug. Several brand name dietary supplements add yohimbe extract as a source of the drug yohimbine.

Yohimbine blocks alpha2-receptors, resulting in an increased blood supply to the corpus cavernosum. Yohimbine increases catecholamine release in peripheral tissues, resulting in the many side effects described later.

Yohimbe

There is evidence that the purified drug yohimbine might be of help for erectile dysfunction.[26-28] Although there have been many claims that the herbal formulation actually works better than the purified drug, there is no evidence that this is the case. In fact, there is no reliable evidence that yohimbe extracts are effective for erectile dysfunction.

Doses of the active constituent yohimbine 15 to 30 mg/day have been used. The appropriate dose of the natural product yohimbe is not known.

SAFETY CONSIDERATIONS/DRUG INTERACTIONS

The purified yohimbine constituent has been safely used in several clinical trials under medical supervision. However, this drug is not appropriate for unsupervised use. Although it may be beneficial for erectile dysfunction, it is associated with many side effects including hypertension, tachycardia, excitation, anxiety, tremor, insomnia, dizziness, gastrointestinal upset, irritability, headache, urinary frequency, fluid retention, nausea and vomiting, and rash.[8] These side effects all pertain to the purified constituent yohimbine. Since yohimbe has not been evaluated in clinical trials, it is not known to what extent the natural product might also cause these side effects.

Yohimbe should be avoided in patients with hypertension, anxiety, psychosis, and benign prostatic hyperplasia. Theoretically, the natural product yohimbe might interact with the same drugs as the purified yohimbine constituent. Yohimbine might increase blood pressure and decrease the effectiveness of antihypertensives and have additive effects with stimulant drugs. Yohimbine might also have additive effects with monoamine oxidase inhibitors.

PATIENT DISCUSSION

Erectile dysfunction is a common condition with a significant effect on the quality of life. Common conditions associated with erectile dysfunction include hypertension, diabetes, ischemic heart disease, hypercholesterolemia, and depression. Eric has co-morbidities and may have both vascular and neurogenic factors contributing to his condition. None of the common products discussed here are likely to offer him acceptable relief. Continued failure can heighten anxiety and worsen his situation. Of these products, l-arginine is the most well supported and the safest. It might be worth a try for some patients. DHEA could be appropriate if there is a documented DHEA deficiency. Since DHEA is associated with erectile dysfunction, it would be reasonable to check DHEA-S levels in patients with idiopathic erectile dysfunction. Levels less than 1.5 micromol/L are considered low.

The evidence for Panax ginseng is too preliminary to recommend this product for erectile dysfunction. There is no evidence to support the use of yohimbe bark. The prescription drug yohimbine may offer some benefit, but sildenafil and other agents in the same class have replaced yohimbine as the agents of choice in the doctor's office due to a higher degree of efficacy and reduced side effect profile.

There are lots of natural products that get promoted for treating erectile dysfunction and for improving sexual performance. Some of these not discussed in this chapter include horny goat weed and damiana. There is no reliable evidence that these products are effective for this use.

Another option for Eric is to purchase a different PDE5 inhibitor with a longer duration of action. Tadalafil (Cialis®) has been approved for a duration of 36 hours; however, there are studies showing efficacy out to 100 hours.[29]

Self-Assessment

1. **Which is FALSE concerning Panax ginseng:**
   a. It may lower blood glucose.
   b. Many side effects have been noted, but it is generally well tolerated.
   c. It can be confidently recommended as suitable therapy for most patients with erectile dysfunction.
   d. Panax ginseng is also called Asian ginseng.

2. **A patient wishes to try DHEA for erectile dysfunction. If you choose to recommend this product, what daily dose would you suggest?**
   a. 500 mg
   b. 100 mg
   c. 50 mg
   d. 2.5 mg

3. **Pick the correct mechanism of action of l-arginine in causing increased blood flow to the penis:**
   a. l-Arginine is converted to nitric oxide, which relaxes the vasculature and causes vasodilation.
   b. l-Arginine is converted to prostacyclin, which relaxes the vasculature and causes vasodilation.
   c. l-Arginine causes a man to feel relaxed, which results in an improved erection
   d. The mechanism of action is unknown.

CHAPTER 5: *Erectile Dysfunction* 59

4. The bark of the evergreen _____tree contains the active constituent _____.
   a. Pine/Paclitaxel
   b. Panax ginseng/Asian ginseng
   c. Yohimbine/Yohimbe
   d. Yohimbe/Yohimbine

5. Which product below is not appropriate for unsupervised use?
   a. Panax ginseng
   b. l-Arginine
   c. Yohimbine
   d. Asian ginseng

6. A patient has diabetes, which is well controlled. He is concerned about not using anything that can cause his blood glucose to drop too low. This has happened to him in the past and he felt dizzy and disoriented. For this reason only, you should counsel him to avoid:
   a. DHEA
   b. Panax ginseng
   c. Yohimbe
   d. Yohimbine

7. Your patient has hypertension. He has started to take DHEA. His blood pressure was well controlled but since he started the DHEA it has been elevated. Which question is most appropriate to ask the patient at this time?
   a. What time of day do you take the DHEA?
   b. How many erections have you had in the past week?
   c. What dose of DHEA are you taking on a daily basis?
   d. All of the above.

8. A patient tells you that his girlfriend's herbalist recommended the product yohimbe for his erectile dysfunction. He says that the herbalist says this is more natural and that mainstream doctors do not like it because it is a cheap, natural product. What do you tell the patient regarding this advice?
   a. Yohimbine, a constituent of yohimbe, may be effective but should not be used over-the-counter due to potentially harmful side effects.
   b. The herbalist is correct.

9. Which of the following natural products is an enzyme inhibitor that could potentially increase levels of other drugs?
   a. Yohimbine
   b. Yohimbe
   c. DHEA
   d. l-Arginine

10. A man presents to your clinic and is interested in taking one of the following products for erectile dysfunction. His health conditions include elevated blood pressure and an elevated PSA. He is being evaluated currently for prostate cancer. Which product do you think has the best supporting data, at present, and may offer modest benefit to this patient?
    a. Yohimbe
    b. Horny goat weed
    c. DHEA
    d. l-Arginine

Answers: 1-c; 2-c; 3-a; 4-d; 5-c; 6-b; 7-c; 8-a; 9-c; 10-d

REFERENCES

1. Tenebaum A, Fisman EZ, Motro M. l-Arginine: rediscovery in progress. *Cardiology.* 1998;90:153–5.

2. Lerman A, Burnett JC Jr, Higano ST, et al. Long-term l-arginine improves small-vessel coronary endothelial function in humans. *Circulation.* 1998;97:2123–8.

3. Adams MR, McCredie R, Jessup W, et al. Oral l-arginine improves endothelium-dependent dilatation and reduces monocyte adhesion to endothelial cells in young men with coronary artery disease. *Atherosclerosis.* 1997;129:261–9.

4. Chen J, Wollman Y, Chernichovsky T, et al. Effect of oral administration of high-dose nitric oxide donor l-arginine in men with organic erectile dysfunction: results of a double-blind, randomized, placebo-controlled study. *BJU Int.* 1999;83:269–73.

5. Klotz T, Mathers MJ, Braun M, et al. Effectiveness of oral l-arginine in first-line treatment of erectile dysfunction in a controlled crossover study. *Urol Int.* 1999;63:220–3.

6. Stanislavov R, Nikolova V. Treatment of erectile dysfunction with pycnogenol and l-arginine. *J Sex Marital Ther.* 2003;29:207–13.

7. Jellin JM, Gregory PJ, Batz F, et al. Natural Medicines Comprehensive Database. Therapeutic Research Faculty. Stockton CA; 2004. Available at www.naturaldatabase.com. Accessed December 27, 2004.

8. Buvat J. Androgen therapy with dehydroepiandrosterone. *World J Urol.* 2003;21:346–55.

9. Arlt W, Haas J, Callies F, et al. Biotransformation of oral dehydroepiandrosterone in elderly men: significant increase in circulating estrogens. *J Clin Endocrinol Metab.* 1999;84:2170–6.

10. Ito TY, Trant AS, Polan ML. A double-blind placebo-controlled study of ArginMax, a nutritional supplement for enhancement of female sexual function. *J Sex Marital Ther.* 2001 Oct-Dec;27(5):54.

11. Baulieu EE, Thomas G, Legrain S, et al. Dehydroepiandrosterone (DHEA), DHEA sulfate, and aging. Contribution of the DHEAge study to a sociobiomedical issue. *Proc Natl Acad Sci U S A.* 2000;97:4279–84.

12. Reiter WJ, Pycha A, Schatzl G, et al. Serum dehydroepiandrosterone sulfate concentrations in men with erectile dysfunction. *Urology.* 2000;55:755–8.

13. Reiter WJ, Schatzl G, Mark I, et al. Dehydroepiandrosterone in the treatment of erectile dysfunction in patients with different organic etiologies. *Urol Res.* 2001;29:278–81.

14. Reiter WJ, Pycha A, Schatzl G, et al. Dehydroepiandrosterone in the treatment of erectile dysfunction: a prospective, double-blind, randomized, placebo-controlled study. *Urology.* 1999;53:590–5.

15. Petri MA, Mease PJ, Merrill JT, et al. Effects of prasterone on disease activity and symptoms in women with active systemic lupus erythematosus. *Arthritis Rheum.* 2004;50:2858–68.

16. Kroboth PD, Salek FS, Pittenger AL, et al. DHEA and DHEA-S: a review. *J Clin Pharmacol.* 1999;39:327–48.

17. Kline MD, Jaggers ED. Mania onset while using dehydroepiandrosterone. *Am J Psychiatry.* 1999;156:971. Letter.

18. Markowitz JS, Carson WH, Jackson CW. Possible dihydroepiandrosterone-induced mania. *Biol Psychiatry.* 1999;45:241–2.

19. Dean CE. Prasterone (DHEA) and mania. *Ann Pharmacother.* 2000;34:1419–22.

20. Calhoun KE, Pommier RF, Muller P, et al. Dehydroepiandrosterone sulfate causes proliferation of estrogen receptor-positive breast cancer cells despite treatment with fulvestrant. *Arch Surg.* 2003;138:879–83.

21. Morris KT, Toth-Fejel S, Schmidt J, et al. High dehydroepiandrosterone-sulfate predicts breast cancer progression during new aromatase inhibitor therapy and stimulates breast cancer cell growth in tissue culture: a renewed role for adrenalectomy. *Surgery.* 2001;130:947–53.

22. Ciolino H, MacDonald C, Memon O, et al. Dehydroepiandrosterone inhibits the expression of carcinogen-activating enzymes in vivo. *Int J Cancer.* 2003;105:321–5.

23. Frye RF, Kroboth PD, Folan MM, et al. Effect of DHEA on CYP3A-mediated metabolism of triazolam. *Clin Pharmacol Ther.* 2000;67:109 (abstract PI-82).

24. Hong B, Ji YH, Hong JH, et al. A double-blind crossover study evaluating the efficacy of Korean red ginseng in patients with erectile dysfunction: a preliminary report. *J Urol.* 2002;168:2070–3.

25. Jellin JM, Gregory PJ, Batz F, et al. Natural Medicines Comprehensive Database. Therapeutic Research Faculty. Stockton CA; 2004. Available at www.naturaldatabase.com. Accessed August 31, 2005.

26. Carey MP, Johnson BT. Effectiveness of yohimbine in the treatment of erectile disorder: four meta-analytic integrations. *Arch Sexual Behavior.* 1996;25:341–60.

27. Ernst E, Pittler MH. Yohimbine for erectile dysfunction: a systematic review and meta-analysis of randomized clinical trials. *J Urol.* 1998;159:433–6.

28. Ashton AK. Yohimbine in the treatment of male erectile dysfunction. *Am J Psychiatry.* 1994;151:1397.

29. Mirone V, Costa P, Damber JE, et al. An evaluation of an alternative dosing regimen with tadalafil, 3 times/week, for men with erectile dysfunction: SURE study in 14 European countries. *Eur Urol.* 2005 Jun;47(6):846–54; discussion 854.

CHAPTER 6

Benign Prostatic Hyperplasia

Karen Shapiro

Men commonly begin to experience nighttime urination as they get older due to benign prostatic hyperplasia (BPH). Similar symptoms can be due to prostate cancer, so physician diagnosis is important. Men can be reluctant to visit their physician, especially if they suspect cancer. It is important not to begin any self-treatment plan until the possibility of cancer has been eliminated.

The two risk factors for BPH are being male and getting older. The prostate grows throughout life and as the gland gets bigger, it may narrow or partially block the urethra, causing urinary problems. BPH requires treatment only if the symptoms are bothersome, and in very severe cases. The American Urological Association (AUA) has a symptom scoring system to help clinicians determine BPH severity. A score between 0 and 7 is considered mild, 8 to 19 is moderate, and 20 to 35 is severe. The higher the score, the more likely it is that the patient will receive treatment. With a moderate score, the physician and patient have the option of starting medication. Alternative products include saw palmetto, the most common herbal medication for BPH. Pygeum, stinging nettle, and lycopene are present in men's prostate

KEY POINTS

- Benign prostatic hyperplasia symptoms may not need to be treated.
- The American Urological Association scoring system helps classify symptom severity.
- Men who present with benign prostatic hyperplasia must see a physician prior to initiating self-treatment.
- Saw palmetto is the best-studied product and is effective for symptom relief.
- Saw palmetto extract must be standardized to contain 85% fatty acids and sterols.
- An effective dose of saw palmetto extract is 160 mg twice daily.
- Pygeum has good data to support efficacy for symptom reduction in benign prostatic hyperplasia.
- Wild pygeum has been overharvested; cultivated sources must be used.
- Stinging nettle is used for this condition; however, efficacy data are insufficient.
- Lycopene is present in many combination products; this is used for prostate cancer prevention.

formulas and men's vitamins. There is no evidence for the use of stinging nettle, but it is presented briefly. Many combination supplements contain lycopene, which is present for prostate cancer prevention, not BPH. A diet rich in lycopene seems to reduce the risk of prostate cancer. Lycopene is a carotenoid and most lycopene in the diet comes from tomatoes.[1] Tomato sauce is a better source than raw tomatoes because heating improves the bioavailability. Supplements may also be useful.

Saw Palmetto

The American dwarf palm tree, or saw palmetto (Serenoa repens), is native to the southeastern United States and grows abundantly in Florida. It is cultivated for commercial use. The berries are used medicinally and are deep red-brown or black and about 1 inch long. In the early 1900s, American druggists and physicians commonly recommended saw palmetto berry fluid extracts for various conditions of the prostate. Today, saw palmetto extract (SPE) is used to treat BPH. The active constituents found in the lipophilic extract include phytosterols, free fatty acids and ethyl esters, glycerides, and beta-sitosterol.

Product	Dosage	Effect	Safety Concerns
Saw palmetto	160 mg twice daily, standardized to 85% fatty acids and sterols; trial period of at least 4 weeks	Effective for reducing BPH symptoms, improving quality of life	• Well tolerated; rare gastrointestinal side effects
Pygeum	100 to 200 mg, taken once daily or in divided doses, of an extract standardized to contain 14% triterpenes and 0.5% n-docosanol	Effective for symptom reduction	• Well tolerated • Wild pygeum has been over-harvested; use cultivated sources
Stinging nettle	Not applicable	No evidence to support use at present	• May decrease effects of warfarin • May decrease blood glucose • May decrease blood pressure • May have additive effects with CNS depressants

Most commercial formulations utilize a liposterolic extract standardized to contain 85% to 95% free fatty acids and sterols. Similar to the pharmacology of finasteride, this extract has been shown to inhibit the activity of 5-α-reductase, an enzyme that converts testosterone to dihydrotestosterone (responsible for promoting prostate tissue growth).[2] In clinical studies, saw palmetto berry extract has not reduced prostate size consistently; therefore, benefits may be derived from multiple mechanisms.

The extract of saw palmetto berries also demonstrates several properties including inhibition of dihydrotestosterone binding at androgen receptors, inhibition of prostaglandin synthesis, and antagonism of alpha1 receptors. However, at clinically recommended doses, the significance of this latter activity is questionable.[3] Numerous randomized, double-blind, clinical trials have demonstrated that saw palmetto berry extract improves symptoms in men with mild-to-moderate BPH.[4] Signs and symptoms that improve include dysuria, nocturia, peak urinary flow rate,

Saw palmetto

PATIENT CASE

Ed

Ed is a 67-year-old retired professional baseball referee. He is married to his wife of 42 years. He does not smoke. He drinks one or two beers several nights a week. Ed is generally healthy and takes no medications. He reports urinary symptoms for 2 years, including increased nighttime urination, difficulty initiating urination, weak flow, and dribbling at the end of urination.

Ed has been to his doctor, who diagnosed benign prostatic hyperplasia (BPH). His American Urological Association symptom score was 8, at the low end of the moderate range. His physician told him that he could give him a prescription for some pills if Ed wanted them. Ed was not quite sure if he needed to start taking prescription pills. His neighbor is using saw palmetto for the same condition and reports that it works. Ed is asking if it is safe and if there are any long-term problems from using this product.

and residual bladder volume. Significant improvements in quality-of-life measurements were also noted.

In a 6-month, randomized, double-blinded trial, saw palmetto berry extract taken 320 mg once daily (Permixon, Pierre Fabre, France) was as effective as finasteride 5 mg/day.[5] Treatment with this extract did not reduce prostate-specific antigen (PSA) levels, and therefore did not mask a lab value used to help detect prostate cancer, and was associated with a very modest reduction in prostate volume. Compared to the α-receptor blockers (e.g., prazosin), saw palmetto berry extract is less effective in improving symptoms.[6]

The recommended dose for BPH is 160 mg twice daily of a liposterolic oil of saw palmetto berry extract standardized to a minimum of 85% fatty acids and sterols. There are significant variations in product quality. In a review by ConsumerLab®, one of the cheapest formulations passed testing and one of the most expensive (five times the cost) did not. A minimum treatment of 4 weeks is required for therapeutic benefits.

Safety Considerations/Drug Interactions

Saw palmetto extract is generally well tolerated with headache and gastrointestinal complaints most commonly reported (e.g., abdominal pain, diarrhea, constipation, and nausea). The frequency of side effects associated with this extract is similar to placebo. The occurrence of erectile dysfunction, ejaculatory disturbance, or altered libido is significantly less than with finasteride. Overall, saw palmetto extract is better tolerated than non-selective α-receptor blockers or finasteride.

Pygeum

Pygeum is an evergreen tree that grows in the mountains of central and southern Africa. Pygeum is also called the African plum tree. Pygeum fruit (called drupe) looks like a cherry. Pygeum is closely related to cherries, plums, and almonds.

Pygeum is an endangered plant. Effort should be made to purchase only pygeum that has been cultivated and not use any that was harvested in the wild. Pygeum extract is made from the bark or branches and removing these can kill the tree. Plantations of trees are being developed in some African countries to meet both international demand and for use in traditional African medicine.

Pygeum has a different mechanism of action than saw palmetto. It does not prevent testosterone conversion. The mechanism seems to be related to suppression of growth factors and anti-inflammatory effects.[7,8] Pygeum may be as effective as saw palmetto for BPH; however, this determination cannot be made presently because, although there are more than 17 controlled trials, none compares pygeum to standard therapies. The study data do support significant symptom improvement. In a controlled, multicenter study with 263 subjects, the group receiving pygeum 50 mg twice daily had significant reductions in American Urological Association parameters.[9] Other trials show similar benefits of about a 20% reduction in symptoms.[10]

Pygeum

The dose used in clinical trials is 100 to 200 mg, taken once daily or in divided doses. The extract should be standardized to contain 14% triterpenes and 0.5% n-docosanol.

SAFETY CONSIDERATIONS/DRUG INTERACTIONS

Pygeum is well tolerated, with occasional minor gastrointestinal complaints. There are no known drug interactions.

STINGING NETTLE

Stinging nettle is native to the United States and Europe. Hikers may have experienced itchy bumps on their skin from rubbing up against these plants. Stinging nettle hairs, or trichomes, grow on the stems and leaves. When brushed, the tip of the trichome breaks off, leaving a sharp needle-like point that injects an acidic irritant into the skin.

Stinging nettle root may have anti-inflammatory effects and an antiproliferative effect on prostate cells.[11] Clinically, there are no current data in human subjects that support the use of this product for BPH. One available study of a combination product that included stinging nettle had disappointing results.[12] This has not prevented stinging nettle from appearing in many prostate and men's vitamin formulas. No dose can be suggested at this time.

SAFETY CONSIDERATIONS/DRUG INTERACTIONS

Stinging nettle can cause gastrointestinal upset. It may decrease the effects of warfarin and could decrease blood glucose levels. It may also increase the effects of antihypertensives and CNS depressants.[11] There is not a lot of clinical experience with this product. It should be used cautiously.

PATIENT DISCUSSION

Ed is a good candidate for saw palmetto. He will need to use the recommended dose, in a tested, standardized product for a reasonable trial of at least 4 weeks. He should monitor improvement with the American Urological Association scoring system. If he does not receive adequate improvement, the product should be discontinued.

Saw palmetto is sometimes marketed in combination with pygeum. The combinations are more expensive, but they can be recommended if the pygeum comes from a sustainable source. Stinging nettle should not be recommended. Lycopene may be present in the product he purchases. This is used for prostate cancer prevention.

Ed should be made aware that alcohol is a prostate irritant and a diuretic. Minimally, he should avoid alcohol consumption 2 to 3 hours before bedtime to help reduce nighttime urination.

SELF-ASSESSMENT

1. Which is TRUE concerning saw palmetto:
    a. The dose for BPH is 160 mg twice daily.
    b. The dose does not need to be standardized.
    c. Efficacy data are of poor quality.
    d. This product has significant drug interactions.

2. Saw palmetto has a mechanism of action similar to finasteride.
    a. True
    b. False

3. Saw palmetto does not increase PSA levels.
    a. True
    b. False

4. Saw palmetto grows abundantly in Florida and is cultivated for commercial use.
    a. True
    b. False

5. Which is TRUE concerning pygeum:
    a. The dose for BPH is 100 to 200 mg/day.
    b. The extract must be standardized.
    c. The extract is effective for symptom reduction.
    d. All of the above.

6. Which is TRUE concerning pygeum:
   a. It is poorly tolerated, with significant gastrointestinal side effects.
   b. There are significant drug interactions.
   c. There is no risk of extinction of the wild plant.
   d. Pygeum is also called the African plum tree.

7. Which of the following foods is rich in lycopene and is used for prostate cancer prevention?
   a. African berry juice
   b. Stinging nettle extract
   c. Tomato sauce
   d. Plum jelly

8. Which of the following products may decrease the effect of warfarin and lower blood glucose levels?
   a. Saw palmetto
   b. Pygeum
   c. Lycopene
   d. Stinging nettle

9. Stinging nettle has good data for symptom reduction in BPH.
   a. True
   b. False

10. A patient should self-monitor prostate symptoms when self-treating. The scoring system is called:
    a. The American Urological Association (AUA) symptom scoring system
    b. The prostate health survey
    c. The urination frequency scoring system
    d. The healthy prostate scoring system

Answers: 1-a; 2-a; 3-a; 4-a; 5-d; 6-d; 7-c; 8-d; 9-b; 10-a

REFERENCES

1. Norrish AE, Jackson RT, Sharpe SJ, et al. Prostate cancer and dietary carotenoids. *Am J Epidemiol.* 2000;151:119-23.

2. Bayne CW, Donnelly F, Ross M, et al. *Serenoa repens* (Permixon®): a 5-α-reductase types I and II inhibitor—new evidence in a coculture model of BPH. *Prostate.* 1999;40:232-41.

3. Goepel M, Dinh L, Mitchell A, et al. Do saw palmetto extracts block human alpha1-adrenoceptor subtypes in vivo? *Prostate.* 2001;46:226-32.

4. Wilt T, Ishani A, Stark G, et al. *Serenoa repens* for benign prostatic hyperplasia. *Cochrane Database Syst Rev.* 2000;(2):CD001423.

5. Carraro JC, Raynaud JP, Koch G, et al. Comparison of phytotherapy (Permixon®) with finasteride in the treatment of benign prostate hyperplasia: a randomized international study of 1,098 patients. *Prostate.* 1996;29:231-40.

6. Grasso M, Montesano A, Buonaguidi A, et al. Comparative effects of alfuzosin versus *Serenoa repens* in the treatment of symptomatic benign prostatic hyperplasia. *Arch Esp Urol.* 1995;48:97-103.

7. Wilt T, Ishani A, MacDonald R, et al. *Pygeum africanum* for benign prostatic hyperplasia. *Cochrane Database Syst Rev.* 2002;CD001044.

8. Levin RM, Das AK. A scientific basis for the therapeutic effects of *Pygeum africanum* and *Serenoa repens*. *Urol Res.* 2000;28:201-9.

9. Barlet A, Albrecht J, Aubert A, et al. Efficacy of *Pygeum africanum* extract in the medical therapy of urination disorders due to benign prostatic hyperplasia: evaluation of objective and subjective parameters. A placebo-controlled double-blind multicenter study. *Wien Klin Wochenschr.* 1990;102:667-73. German.

10. Ishani A, MacDonald R, Nelson D, et al. *Pygeum africanum* for the treatment of patients with benign prostatic hyperplasia: a systematic review and quantitative meta-analysis. *Am J Med.* 2000;109:654-64.

11. Jellin JM, Gregory PJ, Batz F, et al. Natural Medicines Comprehensive Database. Therapeutic Research Faculty. Stockton CA; 2005. Available at www.naturaldatabase.com. Accessed March 15, 2005.

12. Marks L, Partin AW, Epstein JI, et al. Effects of a saw palmetto herbal blend in men with symptomatic benign prostatic hyperplasia. *J Urol.* 2000;163:1451-6.

CHAPTER 7

Osteoporosis

Karen Shapiro

Osteoporosis is a bone-thinning disease that is widespread and causes much unnecessary suffering. Currently, the probability that a 50-year-old will have a hip fracture during their lifetime is 14% for a white female and 5% to 6% for a white male.[1] Common fracture sites are the wrist, hip, and spine. Hip fractures are the most debilitating and are associated with high mortality rates. Vertebral fractures are much more common, begin at an earlier age, and can cause chronic pain.

Osteoporosis is diagnosed when the T-score is –2.50 or lower. Therapy is sometimes used in osteopenia (low bone density) in order to prevent further loss. Patients with previous fractures or gait instabilities and patients taking medications that can cause sedation or confusion are at a much higher risk of falling and will likely receive a prescription agent earlier.[2]

Some prescription agents reduce bone density (e.g., steroids, certain anticonvulsants, and others). When this is the case and the agent cannot be safely discontinued, the baseline calcium and vitamin D must be optimal and, if

KEY POINTS

- Calcium intake is essential for proper bone health throughout the life span.
- Calcium intake should be spread throughout the day in order to maximize absorption.
- Vitamin D is required for optimal calcium absorption.
- Bone loss occurs rapidly at menopause and requires adequate intake of calcium 1200 mg/day and adequate vitamin D.
- "Miracle" calcium products represent false advertising promoted to women who desire to avoid fractures.
- Cheaper, readily available calcium carbonate or citrate formulations can be recommended.
- The evidence for soy as a beneficial agent for osteoporosis is inconclusive.
- Ipriflavone can be recommended for osteoporosis prevention or as adjunctive therapy.
- Patients who need prescription agents for low bone density should be encouraged to use these products; natural products do not offer similar benefit.
- All persons using prescription agents for bone density require adequate calcium and vitamin D.

necessary (such as with chronic oral steroid administration), prophylactic therapy with a prescription bisphosphonate agent may be required.

CALCIUM

The best treatment for osteoporosis is prevention. Historically, calcium intake was about 2000 mg a day or more; current diets average about one-third of that amount.[4,5] Future incidence of osteoporosis is expected to increase dramatically,

Product	Summary of Evidence	Dose	Safety Concerns
Calcium	Builds bone and reduces bone loss; additive effect with vitamin D and prescription agents; required throughout the life-span	**Age, years** — **Daily calcium intake** 1 to 3 .... 500 mg 4 to 8 .... 800 mg 9 to 18 .... 1300 mg 19 to 50 .... 1000 mg 50+ .... 1200 mg Pregnant or lactating and <18 years ...... 1300 mg Pregnant or lactating and 19 to 50 years .. 1000 mg	• Separate doses from levothyroxine, quinolone antibiotics, and tetracycline antibiotics
Vitamin D	Required for calcium absorption and bone mineralization throughout life span	400 IU/day to age 70; 600 IU/day after age 70	• 2000 IU/daily recommended limit, although higher weekly doses are well tolerated
Soy	May increase bone density; data not conclusive	40 to 80 mg soy protein isolate	• Risk for breast cancer unknown; avoid in patients with breast cancer
Ipriflavone	Strongest data of natural products—but not equivalent to prescription agents	200 mg three times daily	• Enzyme inhibitor; can lower concentration of many drugs • May cause subclinical (asymptomatic) lymphocytopenia

largely due to poor calcium intake and low levels of physical activity among children and adolescents. Practitioners should attempt to correct calcium intake in their young patients—this involves stressing the importance of drinking milk rather than sodas and fruit juice. In addition to building strong bones, this will help reduce the incidence of obesity and tooth decay. Calcium has other health benefits, including a significant reduction in PMS symptoms and a reduced risk of colorectal cancer.[6,7] Deficiencies in calcium (and magnesium) can elevate blood pressure.[8,9] Calcium supplementation reduces bone loss in all population groups, with a greater benefit seen in premenopausal women.[10-12]

PATIENT CASES

SANDRA—A PATIENT WITH OSTEOPOROSIS

Sandra is a 61-year-old Caucasian female who just received results of her bone density (DEXA) test today. She has a T-score of -2.6 in the spine and -2.2 in the hip. This is her first bone density test. Her physician wants her to take Actonel® 30 mg once weekly.

Unfortunately, Sandra has reached the maximum number of prescriptions her health insurance will allow and this medication will cost $70 a month. She tells you that she has never taken a calcium supplement and asks if she can use calcium and something else over-the-counter for her bones instead of the Actonel. She is taking amlodipine, atenolol, and hydrochlorothiazide for hypertension, topiramate for seizures, pravastatin for high cholesterol, and sertraline for anxiety. She notes that she is trying hard to improve her bones and has taken up swimming three times weekly.

NIA—A PATIENT WITH NORMAL BONE DENSITY

Nia is a 48-year-old Asian female who received results of her bone density (DEXA) test today. She requested the DEXA exam after she found out that her mother has osteoporosis. She had a T-score of -1.2 in the spine and -0.4 in the hip. Her periods were regular until quite recently.

Nia has been told by her physician that she is in the perimenopause. She is concerned about the negative numbers and asks if she can take anything to help protect her bones. She walks daily. Her diet is rich in vegetables and soy. She does not consume dairy products. She states that milk upsets her stomach and she does not like yogurt or cheese. She thinks she should at least be taking calcium and asks if you recommend miracle coral calcium. She asks if there is a good product to help her bones that can be taken with the calcium. Her medications are loratadine, as needed for allergies, and a daily Centrum® multivitamin.

Which form of calcium should be recommended? Calcium carbonate is cheap, safe, and available in many supplements, including generic formulations. Calcium carbonate has "acid-dependent" absorption and should be taken with food. Calcium citrate is the second most popular supplement. This salt form is more soluble than the carbonate and has good absorption, with or without food. The downside is that calcium citrate supplements are bulkier. The carbonate form contains 40% elemental calcium by weight compared to 21% in the citrate product. This means that one large calcium citrate tablet has about 315 mg calcium versus 500 to 600 mg of calcium in the smaller carbonate tablet. Many elderly patients will find calcium citrate difficult to take without cutting the tablets into smaller pieces. The carbonate forms come in chewable chocolate forms, including Viactiv® and generics. For those who prefer to take calcium on an empty stomach, calcium citrate can be recommended. Some clinicians recommend calcium citrate for elderly patients who may have low gastric acidity or in patients using acid-suppression therapy. It is not clear if this is necessary.[13,14]

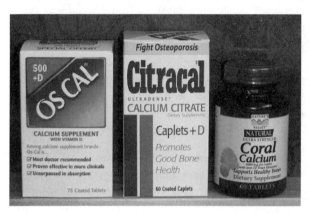

Calcium products

There is calcium, and then there is "miracle" calcium, "high grade coral calcium," "calcium l-threonate" (in the Biocalth® products, which are marketed heavily to Asian women) and microcrystalline hydroxyapatite calcium (in the popular Bone-Up® products), among others. There is no benefit to buying expensive calcium products over cheaper formulations. There is no rationale for most of the added vitamins and minerals in calcium supplements in the majority of patients, with the exceptions of vitamin D and magnesium, if needed.[15,16] Magnesium may benefit patients who are magnesium deficient; however, patients will likely not know if this applies to them.[17] Many of the cheaper supplements come with added magnesium and there is no harm in adding this mineral. If a patient wants other vitamins or minerals, these can be easily obtained from a basic multivitamin and mineral supplement.

Consumers should question the quality of a supplement made by a manufacturer who is willing to make exaggerated or false claims in order to sell a product at an exorbitant cost. This may be especially important where calcium is concerned, since lead levels can exceed safe amounts. Good manufacturing practices include guidelines to test for safe lead levels.

Safety Considerations/Drug Interactions

Calcium absorption is optimal at doses of 500 mg or less at one time. The higher the dose, the lower the percent absorbed. Calcium intake must be spread out through the day in order to maximize absorption. Calcium is generally well tolerated and side effect profiles between different formulations (tablets, liquids, chews) are similar. Stomach upset can occur. A medical myth exists that calcium causes constipation. Typical questions should be asked of patients with constipation, such as fluid and fiber intake, the use of constipating agents, and physical activity routine. Calcium doses should be separated from levothyroxine, quinolone antibiotics, and tetracycline antibiotics.

Vitamin D

It is common knowledge that calcium intake is often low. What is not as well known is that vitamin D deficiency is a widespread problem and must be addressed if the incidence of osteoporosis is to be reduced. The term vitamin D refers to a group of steroid molecules. Vitamin D3, cholecalciferol, is synthesized in the skin from sunlight exposure. Vitamin D can also be obtained from food sources, including vitamin D-fortified dairy products, egg yolks, and fish. The primary effects of vitamin D on bone are to provide the proper balance of calcium and phosphorus absorption and to support bone mineralization. Vitamin D promotes the expression of proteins involved in transporting calcium from the intestinal lumen, across the epithelial cells, and into blood.[18] When vitamin D levels are low and absorption is impaired, the body attempts to compensate by increasing parathyroid hormone secretion. This causes an increase in bone resorption and accelerated bone loss.[19]

Factors Contributing to Low Serum Vitamin D
- Low dietary intake
- Reduced intestinal absorption
- Reduced sunlight exposure
- Reduced conversion ability in aging skin
- Decreased renal function
- Low endogenous synthesis

Source: Reference 19.

Vitamin D plus calcium has been shown to be effective for preventing osteoporosis and decreasing postmenopausal bone loss.[20-22] In one well-designed study of 192 postmenopausal women, supplementation with vitamin D and calcium significantly increased bone mass density, increased serum vitamin D levels, and significantly decreased serum parathyroid hormone levels and biochemical bone remodeling markers.[23] Vitamin D may also help support balance and reduce the risk of falls, although the mechanism for this action is unknown.[24] It has also been shown to decrease fracture risk.[25] Osteomalacia, the adult form of rickets, is observed in severe cases of vitamin D deficiency.

People who are housebound and experience no sunlight exposure are unable to make vitamin D. As adults age, the ability to make vitamin D through the skin decreases. Intestinal absorption may decrease. Other factors, such as renal dysfunction, can contribute to low vitamin D levels. Many of these factors are present in the elderly.

In younger people, 15 minutes of sunlight exposure of the hands, arms, and face two or three times weekly is adequate. Sunscreen use will negate this benefit. People with fairer skin make more vitamin D from sun exposure than those with darker skin. Ironically, women who live in a sunny climate but cover their skin with veils often have vitamin D deficiency and high rates of osteoporosis as they age.[26] The National Academy of Sciences recommends an Adequate Intake (AI) of 400 IU (10 mcg) daily for people 70 years of age and younger and 600 IU (15 mcg) daily for those 70 years of age and older.[27]

Safety Considerations/Drug Interactions

Vitamin D is well tolerated, unless excessive doses are taken and toxicity develops. This does not occur at recommended doses. Potent enzyme inducers, including phenytoin, carbamazepine, phenobarbital, and rifampin decrease the concentration of vitamin D. This is thought to contribute to the increased risk of osteoporosis that occurs with long-term use of these drugs. In patients using these medications chronically, vitamin D (and calcium) must be supplemented.

Soy

Soy products are commonly used for treatment of menopausal symptoms (which include hot flashes, night sweats, and vaginal dryness) and for bone density improvement.[28] There is increased interest in the use of plant products for both indications in light of data from the Women's Health Initiative (WHI), which showed that the risks of long-term use of conventional hormone therapy outweighed the benefits.[29,30]

Soy products contain compounds called phytoestrogens, or plant estrogens, which are much weaker than endogenous estrogen but can mimic some of the same effects. The phytoestrogens present in soy include isoflavones, lignans, and coumestans.[31] From observational data, Asian women who consume more soy isoflavones daily have higher bone density than those with lower daily intake.[32]

The effect of soy (i.e., isoflavone) on bone density has recently been evaluated in a small study of 66 postmenopausal women given 40 g of soy protein (standardized to 2.25 mg of isoflavone per gram of protein) per day during a 6-month trial. The study noted an increase in lumbar spine bone mineral density of 2.2% during the study period. No bone density differences were noted in the hip.[33] So far, there have been no studies documenting the role of soy in preventing fractures or increasing hip bone mineral density. Data from prospective studies have not shown benefit.[34-37] Different study results may be due to using different soy products and the age of the women. Other factors, such as a woman's body weight and calcium intake may also be important.[38] At present, the data suggest but do not confirm, a beneficial effect of soy intake on bone density.

Daily doses used in trials studying bone density improvement range from 40 to 80 mg of soy protein isolate. For soy content in food products, please refer to the chapter on Cholesterol Reduction.

Safety Considerations/Drug Interactions

Soy, taken in food form, is safe for all age groups. Soy is also a heart-healthy food and can help reduce cholesterol levels. Although long-term soy intake may be protective against breast cancer, particularly in premenopausal women where phytoestrogens may block the (stronger) endogenous estrogens in breast tissue, it may not be wise to recommend soy products to women with a history of breast cancer, due to the possibility of estrogen-agonist effects.[39] In postmenopausal women, where endogenous estrogen is low, phytoestrogens may act as estrogen agonists. The majority of breast tumors grow in the presence of estrogen.

In the majority of people, soy as a dietary addition is certainly a healthier protein source than red meat, for both the individual and society. Red clover is another isoflavone-containing plant that has insufficient evidence at this time for benefit in osteoporosis.[40]

Ipriflavone

Many available products marketed for bone density improvement contain ipriflavone, a synthetic agent developed from daidzen and derived from soy. Ipriflavone was developed with the purpose of providing an isoflavone that possessed bone-stimulating effects without providing estrogen-like activity elsewhere in the body. It is a "designer" phytoestrogen. Consequently, ipriflavone does not provide any benefit for relief of menopausal symptoms and is not discussed under that section. Ipriflavone is the natural product with the best data for bone density improvement.

Ipriflavone enhances bone formation, inhibits bone loss, and seems to enhance the bone-protective effects of estrogen.[41] Ipriflavone can level off the rapid bone loss that occurs postmenopause; however, it is important that baseline calcium and vitamin D is optimized.[42,43] There are also data to support the use of ipriflavone in women with osteoporosis.[44,45] Available data involve trials with small numbers of women (typically less than 30) and do not include fracture reduction. This means that no conclusions can be drawn at this time but it can be stated that the benefit appears promising.[46] Patients who need a bisphosphonate will not get adequate bone density improvement with the use of this agent; be sure to counsel women that not all bone density agents are created equal.

The dose of ipriflavone used in the majority of studies is 200 mg three times daily.

SAFETY CONSIDERATIONS/DRUG INTERACTIONS

Ipriflavone is an inhibitor of the cytochrome P450 1A2 and 2C9 enzymes. The concentrations of drugs that are substrates of these enzymes in the body could increase. Selected, but not all, drugs metabolized by 1A2 include propranolol, theophylline, clozapine, olanzapine, cyclobenzaprine, and haloperidol and by 2C9 include diazepam, verapamil, and warfarin.

Ipriflavone may lower lymphocytes in some women. In one study 29 of 234 patients in the ipriflavone group developed subclinical (asymptomatic) lymphocytopenia. Lymphocyte levels returned to normal in 81% of affected patients at 2 years.[47] Lymphocytes should be monitored in future trials and, until more is known, women taking ipriflavone long term should have lymphocytes monitored.

PATIENT DISCUSSION

A PATIENT WITH OSTEOPOROSIS

Sandra has osteoporosis. Her bone density is porous and she may not be able to withstand an impact. Her medications for seizures, hypertension, and depression, along with the disease states themselves, increase her risk for falls. She should be started on an effective prescription bone density agent as soon as possible. This will need to be taken with adequate calcium and vitamin D. Sandra does not take a multivitamin (which usually contain 200 to 400 IU of vitamin D) and should use a calcium supplement with added vitamin D.

Natural products, if she wishes to try them, can be used in addition to the Actonel, but not as a replacement. The pharmacist should work with the patient and the physician to replace two of the individual drugs with a combination drug, which would bring her below her insurance maximum and provide her with coverage. Another option is to provide insurance coverage for the Actonel and have her purchase one of the less expensive drugs.

A PATIENT WITH NORMAL BONE DENSITY

Nia, on the other hand, has good bone health. Her scores are normal. Often when patients are given a report that their bones are "good" they think "Well, that's that" and take no further action. Bone density declines with age and, in this case, the patient does not have a lot in the bank to cushion against future bone loss. The higher the T-scores, the more time someone has until they enter the fracture-risk zone. She requires calcium supplementation.

She will need to consider her total calcium intake at any one time to make sure that the doses are spread out, in order to ensure optimal absorption. The Centrum multivitamin she uses contains 162 mg calcium. Her multivitamin also contains vitamin D. Vitamin D works best when taken with calcium.[48] She is likely getting adequate vitamin D through other sources, including sun exposure and diet. The most important thing for Nia is to optimize her calcium and vitamin D intake. Nia can use ipriflavone, if she wishes, to help protect her bones.

Encourage patients to maintain regular physical activity. Sandra is swimming three times weekly. This should not be discouraged; swimming is great for cardiovascular health and overall flexibility. Yet by itself, swimming is inadequate for healthy bones. Walking and many other forms of exercise can be recommended. In all patients, first check that footware is appropriate, vision is good, gait is uniform, and all factors are in place that will help keep the person upright before recommending a physical activity program.

SELF-ASSESSMENT

1. **The recommended daily dose of calcium in women more than 50 years of age is:**
   a. 800 mg
   b. 1000 mg
   c. 1200 mg
   d. 2000 mg

2. **Calcium citrate is better absorbed than calcium carbonate, but it is also less concentrated.**
   a. True
   b. False

3. Vitamin D intake for women less than 70 years of age is:
   a. 50 IU
   b. 100 IU
   c. 300 IU
   d. 400 IU

4. Which is FALSE concerning calcium supplementation:
   a. Calcium supplementation can significantly decrease PMS symptoms.
   b. Calcium carbonate is the active ingredient in Oscal®.
   c. Constipation is a common side effect.
   d. Doses should be spread throughout the day.

5. A woman purchases a calcium product with added magnesium. This may be useful in women who are constipated or are magnesium-deficient.
   a. True
   b. False

6. A 35-year-old female patient is interested in reducing her risk of osteoporosis. Which of the following recommendations is appropriate:
   a. She requires adequate calcium intake.
   b. She requires adequate vitamin D.
   c. She should be engaged in weight-bearing physical activity.
   d. All of the above.

7. A woman has osteoporosis and was prescribed Fosamax®. She tells you that she had gas from this product and stopped taking it. She asks if she can use ipriflavone supplements as a replacement. Her total daily calcium and vitamin D intake is acceptable. You reply that this is fine, based on the reports of the GI upset. This advice is acceptable.
   a. True
   b. False

8. The product with the best data for osteoporosis prevention in postmenopausal women is:
   a. Ginkgo biloba
   b. Vitamin E
   c. Soy
   d. Ipriflavone

9. You wish to recommend ipriflavone for a patient. A reasonable dose, based on trial data, is:
   a. 200 mg once daily
   b. 200 mg twice daily
   c. 200 mg three times daily
   d. None of the above

10. Ipriflavone can inhibit CYP enzymes. This means that the concentration of other drugs metabolized by these enzymes, including certain psychiatric medications and theophylline, would be expected to:
    a. Increase
    b. Decrease
    c. Remain the same

Answers: 1-c; 2-a; 3-d; 4-c; 5-a; 6-d; 7-b; 8-d; 9-c; 10-a

REFERENCES

1. NIH Consensus Development Panel on Osteoporosis Prevention, Diagnosis, and Therapy. Osteoporosis prevention, diagnosis, and therapy. JAMA. 2001;285(6):785-95. Review.

2. Heaney RP. Bone mass, bone fragility, and the decision to treat. JAMA. 1998;280(24):2119-20.

3. Available at www.nal.usda.gov/fnic/etext/000105.html. Accessed September 3, 2005.

4. Eaton SB, Nelson DA. Calcium in evolutionary perspective. Am J Clin Nutr. 1991;54:281S-75.

5. Lindsey R, Nieves J. Milk and bones. BMJ. 1994;308:930-1.

6. Thys-Jacobs S, Starkey P, Bernstein D, et al. Calcium carbonate and the premenstrual syndrome: effects on premenstrual and menstrual symptoms. Premenstrual Syndrome Study Group. Am J Obstet Gynecol. 1998 Aug;179(2):444-52.

7. Baron JA, Beach M, Mandel JS, et al. Calcium supplements for the prevention of colorectal adenomas. Calcium Polyp Prevention Study Group. N Engl J Med. 1999;340:101-7.

8. Griffith LE, Guyatt GH, Cook RJ, et al. The influence of dietary and nondietary calcium supplementation on blood pressure: an updated metaanalysis of randomized controlled trials. Am J Hypertens. 1999;12(1 Pt 1):84-92.

9. Witteman JC, Grobbee DE, Derkx FH, et al. Reduction of blood pressure with oral magnesium supplementation in women with mild to moderate hypertension. Am J Clin Nutr. 1994;60:129-35.

10. Cumming RG. Calcium intake and bone mass: a quantitative review of the evidence. *Calcif Tissue Int.* 1990;47:194-201.

11. Prince RL. Diet and the prevention of osteoporotic fractures. *N Engl J Med.* 1997;337:701-702. Editorial.

12. Dawson-Hughes B, Harris SS, Krall EA, et al. Effect of withdrawal of calcium and vitamin D supplements on bone mass in elderly men and women. *Am J Clin Nutr.* 2000;72:745-750.

13. Heaney RP, Dowell MS, Barger-Lux MJ. Absorption of calcium as the carbonate and citrate salts, with some observations on method. *Osteoporos Int.* 1999;9(1):19-23.

14. Heller HJ, Stewart A, Haynes S, et al. Pharmacokinetics of calcium absorption from two commercial calcium supplements. *J Clin Pharmacol.* 1999;39:1151-4.

15. Levenson DI, Bockman RS. A review of calcium preparations [published erratum appears in Nutr Rev. 1994;52:364]. *Nutr Rev.* 1994;52:221-32.

16. Gueguen L, Pointillart A. The bioavailability of dietary calcium. *J Am Coll Nutr.* 2000;19:119S-136S. Review.

17. Gur A, Colpan L, Nas K, et al. The role of trace minerals in the pathogenesis of postmenopausal osteoporosis and a new effect of calcitonin. *J Bone Miner Metab.* 2002;20:39-43.

18. Norman AW. Intestinal calcium absorption: a vitamin D-hormone-mediated adaptive response. *Am J Clin Nutr.* 1990;51:290-300.

19. Reginster J-Y. The high prevalence of inadequate serum vitamin D levels and implications for bone health. *Curr Med Res Opin.* 2005;21(4):579-85.

20. Dawson-Hughes B, Harris SS, Krall EA, et al. Effect of calcium and vitamin D supplementation on bone density in men and women 65 years of age or older. *N Engl J Med.* 1997;337:670-6.

21. Dawson-Hughes B, Dallal GE, Krall EA, et al. Effect of vitamin D supplementation on wintertime and overall bone loss in healthy postmenopausal women. *Ann Intern Med.* 1991;115:505-12.

22. Hunter D, Major P, Arden N, et al. A randomized controlled trial of vitamin D supplementation on preventing postmenopausal bone loss and modifying bone metabolism using identical twin pairs. *J Bone Miner Res.* 2000;15:2276-83.

23. Grados F, Brazier M, Kamel S, et al. Prediction of bone mass density variation by bone remodeling markers in postmenopausal women with vitamin D insufficiency treated with calcium and vitamin D supplementation. *J Clin Endocrinol Metab.* 2003;88:5175-9.

24. Pfeifer M, Begerow B, Minne H, et al. Effects of a short-term vitamin D and calcium supplementation on body sway and secondary hyperparathyroidism in elderly women. *J Bone Miner Res.* 2000;15:1113-8.

25. Minne HW, Pfeifer M, Begerow B, et al. Vitamin D and calcium supplementation reduces falls in elderly women via improvement of body sway and normalization of blood pressure: a prospective, randomized, and double-blind study. *Abstracts World Congress on Osteoporosis;*2000.

26. Guzel R, Kozanoglu E, Guler-Uysal F, Soyupak S, et al. Vitamin D status and bone mineral density of veiled and unveiled Turkish women. *J Womens Health Gend Based Med.* 2001 Oct;10(8):765-70.

27. New Dietary Recommendations. Available at www4.nationalacademies.org/news.nsf/isbn/0309072905?OpenDocument. Accessed August 21, 2005.

28. Mahady GB, Parrot J, Lee C, et al. Botanical dietary supplement use in peri- and postmenopausal women. *Menopause.* 2003 Jan-Feb;10(1):65-72.

29. Rossouw JE, Anderson GL, Prentice RL, et al. Risks and benefits of estrogen plus progestin in healthy postmenopausal women: principal results from the Women's Health Initiative randomized controlled trial. *JAMA.* 2002;288(3):321-33.

30. Anderson GL, Limacher M, Assaf AR, et al. Effects of conjugated equine estrogen in postmenopausal women with hysterectomy: the Women's Health Initiative randomized controlled trial. *JAMA.* 2004;291(14):1701-12.

31. Krebs EE, Ensrud KE, MacDonald R, et al. Phytoestrogens for treatment of menopausal symptoms: a systematic review. *J Obstet Gynecol.* 2004;104:824-36.

32. Somekawa Y, Chiguchi M, Ishibashi T, et al. Soy intake related to menopausal symptoms, serum lipids, and bone mineral density in postmenopausal Japanese women. *Obstet Gynecol.* 2001;97:109-15.

33. Alekel DL, St. Germain A, Peterson CT, et al. Isoflavone-rich soy protein isolate attenuates bone loss in the lumbar spine of perimenopausal women. *Am J Clin Nutr.* 2000;72:844-52.

34. Setchell KDR, Lydeking-Olsen E. Dietary phytoestrogens and their effect on bone: evidence from in vitro and in vivo, human observational, and dietary intervention studies. *Am J Clin Nutr.* 2003;78(3 Suppl):593S-609S.

35. Arjmandi BH, Khalil DA, Smith BJ, et al. Soy protein has a greater effect on bone in postmenopausal women not on hormone replacement therapy, as evidenced by reducing bone resorption and urinary calcium excretion. *J Clin Endocrinol Metab.* 2003;88:1048-54.

36. Kreijkamp-Kaspers S, Kok L, Grobbee DE, et al. Effect of soy protein containing isoflavones on cognitive function, bone mineral density, and plasma lipids in postmenopausal women: a randomized controlled trial. *JAMA.* 2004;292:65-74.

37. Potter SM. Soy protein and isoflavones: their effects on blood lipids and bone density in postmenopausal women. *Am J Clin Nutr.* 1998;68(6 Suppl):1375S-9S.

38. Chen YM, Ho SC, Lam SS, et al. Beneficial effect of soy isoflavones on bone mineral content was modified by years since menopause, body weight, and calcium intake: a double-blind, randomized, controlled trial. *Menopause.* 2004;11:246-54.

39. Murkies A, Dalais FS, Briganti EM, et al. Phytoestrogens and breast cancer in postmenopausal women: a case control study. *Menopause.* 2000;7:289-96.

40. Atkinson C, Compston JE, Robins SP, et al. The effects of isoflavone phytoestrogens on bone: preliminary results from a large randomized, controlled trial. Endocrine Society 82nd Annual Meeting, Toronto, Canada. June 21-4, 2000; Abstract 196.

41. Melis GB, Paoletti AM, Cagnacci A, et al. Lack of any estrogenic effect of ipriflavone in postmenopausal women. *J Endocrinol Invest.* 1992;15(10):755-61.

42. Gennari C, Agnusdei D, Crepaldi G, et al. Effect of ipriflavone—a synthetic derivative of natural isoflavones—on bone mass loss in the early years after menopause. *Menopause.* 1998 Spring;5(1):9-15.

43. Agnusdei D, Gennari C, Bufalino L. Prevention of early postmenopausal bone loss using low doses of conjugated estrogens and the non-hormonal, bone-active drug ipriflavone. *Osteoporos Int.* 1995;5(6):462-6.

44. Agnusdei D, Bufalino L. Efficacy of ipriflavone in established osteoporosis and long-term safety. *Calcif Tissue Int.* 1997;61(Suppl 1):S23-7 [includes review].

45. Adami S, Bufalino L, Cervetti R, et al. Ipriflavone prevents radial bone loss in postmenopausal women with low bone mass over 2 years. *Osteoporos Int.* 1997;7:119-25.

46. Agnusdei D, Crepaldi G, Isaia G, et al. A double blind, placebo-controlled trial of ipriflavone for prevention of postmenopausal spinal bone loss. *Calcif Tissue Int.* 1997;61:142-7.

47. Alexandersen P, Toussaint A, Christiansen C, et al. Ipriflavone in the treatment of postmenopausal osteoporosis: a randomized controlled trial. *JAMA.* 2001;285:1482-8.

48. Mortensen L, Charles P. Bioavailability of calcium supplements and the effect of vitamin D: comparisons between milk, calcium carbonate, and calcium carbonate plus vitamin D. *Am J Clin Nutr.* 1996 Mar;63(3):354-7.

CHAPTER 8

Menopause

Karen Shapiro

Hot flashes (or flushes) are experienced by up to 85% of women with natural menopause. These are thought to be due to the decline in estradiol production by the ovaries, resulting in dysregulation of the central thermoregulatory (temperature) zone. They are most common 1 to 2 years after the last menstrual period and gradually decrease after that, although they last longer in some women. Other symptoms include sweating, the body's attempt to release heat, and mood changes, which can be caused by lack of sleep or can be more complicated. Hot flashes and night sweats will eventually go away by themselves, but supportive therapy is available.

BLACK COHOSH

Black cohosh (*Actaea racemosa*) is discussed first as it is widely known and perhaps the most studied herbal used for menopausal symptoms. Black cohosh is native to North America and was traditionally used by Native Americans for women's health concerns.

The majority of clinical trials show benefit for symptom reduction; however, a conclusion on efficacy is difficult because most trials were open, participants were not blinded and long-term effects were not studied. Several studies were funded by the product manufacturer.[1-4] The American College of Obstetricians and Gynecologists (ACOG) guidelines on the use of botanicals for the manage-

KEY POINTS

- Black cohosh, and possibly soy, can be tried for menopausal symptoms.
- Dong quai does not provide symptom relief when used as a single agent.
- Dong quai has a serious herb-drug interaction with warfarin and can increase bleeding risk.
- Data do not support the use of red clover for menopausal symptoms.
- Data do not support the use of evening primrose oil for menopausal symptoms.
- Bioidentical hormone replacement therapy may provide relief for some women.
- Until long-term study data are available, the use of bioidentical hormone replacement therapy should follow safety precautions similar to those for conventional hormone therapy.

ment of menopausal symptoms support the use of black cohosh for up to 6 months, especially in treating the symptoms of sleep and mood disturbance and hot flushes [evidence level C, consensus/expert opinion].[5] The logic behind ACOG's position may be that short-term use of black cohosh for most women is safe and may provide benefit. If it does not, prescription remedies can be tried next.

Initially, benefit from black cohosh use was thought to be due to estrogenic effects; however, investigation of various extracts found no estrogenic com-

Black cohosh

Product	Dosage	Effect	Safety Concerns
Black cohosh	20-mg tablet taken twice daily (Remifemin®); 4-to-12 week trial	May provide benefit to some women; conflicting data	• Avoid in history of breast cancer • Monitor for hepatotoxicity (case reports)
Dong quai	N/A	No evidence for efficacy when used alone; used in combination in traditional Chinese medicine	• Anticoagulant effect; do not use in patients taking warfarin or in those at risk for bleeding
Soy	Soy protein isolate 60 g/day	May provide benefit to some women; conflicting data	• Risk for breast cancer unknown; avoid in patients with breast cancer
Evening primrose oil	N/A	No evidence for efficacy	• High doses may increase bleeding risk
Bioidentical hormone replacement therapy	Patient specific	Unknown	• Long-term risks unknown; consider same risks as conventional hormone therapy

ponents present. This conclusion has helped allay fears on the risk of breast cancer from black cohosh use, which is now thought to be negligible.[6-9] However, this product should not be recommended for women with a history of breast cancer. First, this product does not appear to offer symptom relief in women with breast cancer, whether or not they are taking tamoxifen.[10] Second, the complete safety profile in regard to estrogen receptors is not resolved. There is some laboratory evidence that black cohosh might affect estrogen receptors, although this is not thought to be the mechanism of action for menopausal symptom relief.[11] The mechanism may be attributable to potentially active ingredients which have been identified, including glycosides, alkaloids, flavonoids, and tannins.[12]

The majority of clinical studies that showed benefit used a standardized extract available in the product Remifemin® (GlaxoSmithKline), although the formulation used in this product (and in studies) has changed over time. Currently, this product is standardized to contain 1 mg of terpene glycosides per 20-mg tablet.[13] It is reasonable to recommend this product. The daily dose is one 20-mg tablet taken twice daily.

Safety Considerations/Drug Interactions

There are case reports of liver damage with black cohosh use. Whether or not these cases were caused by black cohosh is unknown. To be cautious, patients should be counseled to watch for signs of liver damage. Patients using other

PATIENT CASE

Laura

Laura is a 48-year-old white female who presents with complaints of hot flashes, night sweats, and "moody" behavior. Her hot flashes occur several times daily. She gets sweaty at night and feels as if she has not slept well for the past few weeks. She states that she is feeling miserable. She tried taking dong quai, which a friend recommended, and is not sure if it helped. She is interested in taking something natural but only if it offers noticeable symptom relief.

Laura's last menstrual period was 13 months ago. Her T-scores are –0.2 at the hip and –0.4 in the spine. She takes no prescription or over-the-counter medications. Her mother is alive and well at 76. Her father died of prostate cancer at age 74. Her height is 5'4", weight 124 pounds. Her cholesterol panel is normal (LDL = 88 mg/dL) and BP is 110/68 mmHg. She has no brothers or sisters. She is married with one child. She does not use tobacco. She drinks alcohol occasionally at social events. She walks daily for at least half an hour with her two dogs and reports that brisk walking helps somewhat, but not enough.

potentially hepatotoxic agents (methotrexate, isoniazid, amiodarone, etc.) should not use black cohosh. There are no known drug interactions. Side effects are rare but can infrequently include stomach upset and, rarely, dizziness.

Since many women report significant symptom relief with the use of black cohosh and it has a good safety profile (with the unresolved issue of hepatotoxicity as an exception), it is often recommended. Symptom relief is not immediate and women should be counseled that they may not receive full benefit for 4 to 12 weeks. If there is no noticeable symptom relief after a reasonable trial, the black cohosh should be discontinued. Women should be counseled that it does not provide symptom relief for all who try it.

Soy

Soy beans

Menopausal women living in Asian countries experience a much lower incidence of menopausal symptoms. The reasons for this are unknown and could be multifactorial. One factor may be due to dietary intake of phytoestrogens that are contained in soy products. In a double-blind, placebo-controlled study, hot flashes were reduced 45% in women consuming 60 g of soy protein isolate daily.[14] The placebo group had a 30% reduction. The difference between the soy and placebo groups was significant. Other studies have shown significant decreases in symptoms in the soy group versus placebo.[15,16] However, not all studies show benefit.[17] Reasons for conflicting trial data may include differences in the ages of participants and variability in soy products.

Soy, like black cohosh, does not work for everyone. Some women have symptoms that will require stronger prescription therapy. A reasonable dose would be 60 g of soy protein isolate daily. (Soy content in food products is covered in the chapter on *Cholesterol Reduction*.)

Red clover, another phytoestrogen included in the popular product Promensil®, has not been shown to be effective for menopausal symptoms.[18,19]

Safety Considerations/Drug Interactions

Soy, taken in food form, is safe for all age groups. Soy is also a heart-healthy food and can help reduce cholesterol levels, which can be of particular importance at this time. Although long-term soy intake may be protective against breast cancer, particularly in premenopausal women where phytoestrogens may block the (stronger) endogenous estrogens in breast tissue, it may not be wise to recommend soy products to women with a history of breast cancer, due to the possibility of estrogen-agonist effects.[20] In postmenopausal women, where endogenous estrogen is low, phytoestrogens may act as estrogen agonists. The majority of breast tumors grow in the presence of estrogen.

Dong Quai

Dong quai root is used in traditional Chinese medicine in concert with other herbs for gynecological problems. Dong quai is second only to ginseng in reputation in traditional Chinese medicine and is considered the ultimate, all-purpose woman's herbal. Dong quai root has anticoagulant, vasodilatory, and antispasmodic effects.

There is one good study that investigated dong quai's usefulness for menopausal symptoms.[21] In this double-blinded, randomized trial of 71 women receiving dong quai or placebo for 24 weeks, no benefit was seen in symptom reduction. Proponents of traditional Chinese medicine would argue that this study proved nothing, since dong quai is not expected to have benefit when used alone for this purpose. However, studying agents in combination is not typically done and without these data it is impossible to discuss benefit, except anecdotally.

It is reasonable to conclude that dong quai, when used alone, is no more useful than placebo in treating menopausal symptoms. Nonetheless, it is marketed alone and with other agents such as black cohosh.

Safety Considerations/Drug Interactions

Dong quai has coumarin components and has a serious interaction with warfarin, which can result in an elevated international normalized ratio and increased bleeding risk.[22,23] It should not be used by patients taking warfarin or by any patient with a risk of bleeding or by persons taking agents that increase bleeding risk.

Evening Primrose Oil

Evening primrose oil is made from the seeds of the evening primrose plant, named because it blooms at dusk. Evening primrose oil is another agent widely used for menopausal symptoms with little evidence of therapeutic benefit. However, this oil may have other benefits. The seeds are rich in γ-linolenic acid, an essential omega-6 fatty acid. An essential fatty acid is one that must be supplied by the diet, since the body cannot manufacture it or cannot manufacture enough of it. γ-Linolenic acid is a polyunsaturated fat, which is heart-healthy (in contrast to saturated fats from meats and many processed foods, which can contribute to cardiovascular disease). Consumption of γ-linolenic acid is thought to increase production of prostaglandins that may help suppress inflammation.[24] Evening primrose oil is being studied for use in inflammatory conditions such as rheumatoid arthritis.[24,25] It is also used for breast pain (mastalgia) associated with the menstrual cycle.[26] The flowers are edible and the petals, which come in yellow, white, rose, and violet, are pretty additions to salads.

Safety Considerations/Drug Interactions

Like fish oils, high doses of evening primrose oil may contribute to an increased bleeding risk and should be used with caution in patients with a history of bleeding or in patients taking other agents that increase bleeding risk.

Bioidentical Hormone Replacement Therapy

Many women and their physicians believe that bioidentical hormone replacement therapy (BHRT) is a more natural and safer approach to menopause than prescription agents. Bioidentical hormones are identical in chemical structure to those found in a premenopausal woman's body. Premenopausal women make three estrogens: estradiol, estriol, and estrone. Postmenopausal women mostly have estrone. In terms of being "natural," BHRT would actually be termed synthetic or semisynthetic by a pharmacologist since the hormone components are manufactured in a laboratory, often, but not always, from plant-derived material. Premarin®, in contrast, is made from pregnant mare urine and is actually a "natural" product in the sense that the components are not altered. BHRT is, however, bioidentical to a female human and not a female horse.

BHRT typically involves formulations customized to treat a woman's hormone levels and symptoms. A woman can also use preformulated, bioidentical products, including prescription agents such as Estrace® (for estradiol) and Prometrium® (for progesterone).

BHRT is big business. Among other proponents, Suzanne Somers has a book on BHRT called *The Sexy Years: Discover the Hormone Connection—The Secret to Fabulous Sex, Great Health, and Vitality, for Women and Men.* Who wouldn't want that? Dr. Andrew Weil, a physician who has become a leader of alternative therapies in the United States, recommends BHRT for women who have menopausal symptoms, but soundly adds that women trying this approach should limit their intake to low doses for a limited time period.[27] BHRT has been given a boost by results from the Women's Health Initiative trial, which used the conjugated equine hormones present in Premarin and medroxyprogesterone acetate, a synthetic progestin. The WHI trial data invalidated many of the traditional uses of Premarin and medroxyprogesterone acetate, including benefit for cardiovascular disease risk reduction.

Are bioidentical hormones better tolerated, more efficacious, and safer than non-human hormones? In regard to tolerability, clinicians and patients report that they typically are associated with fewer side effects. If this is accurate, it may be because compounded products are developed to suit each woman's hormone needs, unlike standardized drugs. With set dosages, such as with standardized drugs, excess drug may be available to react with more of the "wrong" receptors, creating a higher incidence of side effects.

Efficacy is difficult to determine for a number of reasons. First, if a person pays for a product formulated specifically for them, one would expect a significant placebo response. The second problem is that trial data are conflicting and there are few studies to compare. In one randomized, placebo-controlled trial of healthy, postmenopausal women receiving 20 mg of progesterone cream or placebo for hot flashes, there was significant improvement in symptoms in the treatment group.[28] In another trial, also a randomized, placebo-controlled trial using 32 mg of progesterone daily, no improvement in hot flashes was seen.[29]

Trying to compare these trials is difficult because age from menopause varied within the groups, as did time of analysis and dosage. In one of the trials, the participants received vitamin and mineral supplementation which may have made a difference. What is clear is that larger, well-designed trials looking at both estrogens and progestins used in bioidentical formulations are necessary. This may be wishful thinking because such studies are unlikely to be funded.

SAFETY CONSIDERATIONS/DRUG INTERACTIONS

Any hormone replacement product should be used at the lowest effective dose for the shortest time possible. It is reasonable to exclude women with cardiovascular risk from using BHRT therapy, based on data from the WHI trial. While some practitioners of BHRT would argue that this is unnecessary as the prescription formulations caused the increase in cardiovascular risk, not the BHRT products, it is safest to consider these risks until more is known. Many times, BHRT therapy is taken for years. If there is risk, then the time on therapy should be limited.

Women might also consider that other proponents of natural health consider it unnatural for a postmenopausal woman to have estradiol or progesterone at all, both of which are not "naturally" produced at this time in a woman's life. All women taking estrogen systemically who have a uterus will need progestin opposition to reduce the risk of endometrial cancer. And, all contraindications (including breast cancer and a history of clotting disorders) and precautions to conventional hormone use should remain the same for those taking these hormones in alternative formulations.

PATIENT DISCUSSION

Laura says that her symptoms are considerably bothersome. She can try a natural product but may need something stronger. A reasonable product to try is black cohosh. She can also try to increase her soy intake, either through food or supplements. Black cohosh, with or without soy, should be given a reasonable trial of 4 to12 weeks, if she can tolerate the symptoms that long.

Based on results from the WHI and HERS trial data, Laura may be offered estrogen, or low-dose estrogen, for a limited time period. This is acceptable in women with low cardiac risk.[30] She will, however, have to be presented with the risks and benefits of using hormone therapy. There are other prescription agents that can be helpful and might be useful in her case, including some medications traditionally used for depression, hypertension, and seizures. She should be counseled to avoid hot flash triggers (alcohol, caffeine, hot or spicy foods, stress, and hot drinks.) She is living a healthy life, but if she was overweight, smoking, not exercising, or eating poorly, the clinician could recommend that she also attempt lifestyle changes.

BHRT may offer her relief and may also be worth trying as her cardiovascular disease risk is low. If she chooses to try BHRT, she should use the lowest possible dose for the shortest possible time. Dong quai, as a single agent, red clover, and evening primrose oil are not suitable options.

Laura is doing everything right—exercising, not smoking, and drinking moderately. Her physician may consider a prescription agent for her symptoms if these alternatives do not offer acceptable symptom relief. In addition, women presenting with these complaints present an opportunity to discuss prevention of osteoporosis and cardiovascular disease.

Self-Assessment

1. Which is TRUE concerning black cohosh:
   a. It was used by Native Americans for menopausal symptoms.
   b. It takes 4 to 12 weeks for full benefit.
   c. It does not work for all women.
   d. All of the above.

2. Which is TRUE concerning black cohosh:
   a. The risk of breast cancer is considered small.
   b. Do not recommend in women with a history of breast cancer.
   c. There have been case reports of liver damage.
   d. All of the above.

3. Why has soy been studied for use as a treatment option for menopausal symptoms?
   a. Persons who consume diets rich in soy have less menopausal symptoms.
   b. Isoflavones present in soy can act as very weak estrogens.
   c. Soy contains the female human sex hormones estradiol and estrone.
   d. a and b.

4. Which is TRUE concerning dong quai:
   a. Dong quai is used in concert with other agents in traditional Chinese medicine.
   b. Dong quai provides symptom relief as a stand-alone agent for menopausal symptoms.
   c. Dong quai has no serious drug interactions.
   d. Dong quai can be safely recommended for all women.

5. Evening primrose oil is rich in:
   a. EPA
   b. DHA
   c. GLA
   d. ALA

6. Evening primrose oil may not be useful for hot flashes, but it is being studied for:
   a. Inflammatory disorders
   b. Osteoporosis
   c. Glaucoma
   d. Hirsutism

7. A woman wishes to try bioidentical hormone replacement therapy (BHRT). She has no cardiovascular risk factors and plans to use BHRT for the shortest possible time. She has no history or known risk of breast cancer. A trial period of BHRT is:
   a. Reasonable
   b. Not reasonable

8. A female patient has high blood pressure. She complains of occasional chest pain. She is skinny and smokes five or six cigarettes daily. She wishes to try BHRT for her menopausal symptoms. A trial period of BHRT is:
   a. Reasonable
   b. Not reasonable

9. The long-term health consequences of BHRT are well studied and considered low risk for most women:
   a. True
   b. False

10. Which is TRUE concerning hot flashes:
    a. Eventually they will go away by themselves.
    b. They can sometimes be very disruptive, including causing sleep disturbances.
    c. Severe hot flashes can occur at night and are called night sweats.
    d. All of the above.

Answers: 1-d; 2-d; 3-d; 4-a; 5-c; 6-a; 7-a; 8-b; 9-b; 10-d

REFERENCES

1. Wuttke W, Seidlova-Wuttke D, Gorkow C. The Cimicifuga preparation BNO 1055 vs. conjugated estrogens in a double-blind placebo-controlled study: effects on menopause symptoms and bone markers. *Maturitas*. 2003;44:S67-77.

2. Stoll W. Phytopharmacon influences atrophic vaginal epithelium-double blind study-Cimicifuga vs. estrogenic substances. *Therapeutikon*. 1987;1:23-31. German.

3. Liske E, Wustenberg P. Therapy of climacteric complaints with Cimicifuga racemosa: herbal medicine with clinically proven evidence. *Menopause*. 1998;5:250. Abstract.

4. McKenna DJ, Jones K, Humphrey S, et al. Black cohosh: efficacy, safety, and use in clinical and preclinical applications. *Altern Ther Health Med*. 2001;7:93-100.

5. ACOG Practice Bulletin. Clinical Management Guidelines for Obstetrician-Gynecologists. Use of botanicals for management of menopausal symptoms. *Obstet Gynecol*. 2001;97(Suppl): 1-11.

6. Kennelly EJ, Baggett S, Nuntanakorn P, et al. Analysis of thirteen populations of black cohosh for formononetin. *Phytomedicine.* 2002 Jul;9(5):461-7.

7. Kronenberg F, Fugh-Berman A. Complementary and alternative medicine for menopausal symptoms: a review of randomized, controlled trials. *Ann Intern Med.* 2002 Nov 19;137(10):805-13. Review.

8. Freudenstein J, Dasenbrock C, Nisslein T. Lack of promotion of estrogen-dependent mammary gland tumors in vivo by an isopropanolic Cimicifuga racemosa extract. *Cancer Res.* 2002;62:3448-52.

9. Bodinet C, Freudenstein J. Influence of marketed herbal menopause preparations on MCF-7 cell proliferation. *Menopause.* 2004;11:281-9.

10. Jacobson JS, Troxel AB, Evans J, et al. Randomized trial of black cohosh for the treatment of hot flashes among women with a history of breast cancer. *J Clin Oncol.* 2001;19:2739-45.

11. Davis VL, Jayo MJ, Hardy ML, et al. Effects of black cohosh on mammary tumor development and progression in MMTV-neu transgenic mice. Paper presented at 94th Annual Meeting of the American Association for Cancer Research, July 11-14, 2003; Washington, DC; Abstract R910.

12. Kligler B. Black cohosh. *Am Fam Physician.* 2003 Jul 1;68(1):114-6.

13. Remifemin product information. Available at www.remifemin.com. Accessed December 29, 2004.

14. Albertazzi P, Pansini F, Bonaccorsi G, et al. The effect of dietary soy supplementation on hot flushes. *Obstet Gynecol.* 1998;91:6-11.

15. Burke GL, Legault C, Anthony M, et al. Soy protein and isoflavone effects on vasomotor symptoms in peri- and postmenopausal women: the Soy Estrogen Alternative Study. *Menopause.* 2003;10:147-53.

16. Han KK, Soares JM Jr, Haidar MA, et al. Benefits of soy isoflavone therapeutic regimen on menopausal symptoms. *Obstet Gynecol.* 2002;99:389-94.

17. Secreto G, Chiechi LM, Amadori A, et al. Soy isoflavones and melatonin for the relief of climacteric symptoms: a multicenter, double-blind, randomized study. *Maturitas.* 2004;47:11-20.

18. Tice JA, Ettinger B, Ensrud K, et al. Phytoestrogen supplements for the treatment of hot flashes: the Isoflavone Clover Extract (ICE) study: a randomized controlled trial. *JAMA.* 2003;290:207-14.

19. Knight DC, Howes JB, Eden JA. The effect of Promensil, an isoflavone extract, on menopausal symptoms. *Climacteric.* 1999;2:79-84.

20. Murkies A, Dalais FS, Briganti EM, et al. Phytoestrogens and breast cancer in postmenopausal women: a case control study. *Menopause.* 2000;7:289-96.

21. Hirata JD, Swiersz LM, Zell B, et al. Does dong quai have estrogenic effects in postmenopausal women? A double-blind, placebo-controlled trial. *Fertil Steril.* 1997;68:981-6.

22. Page RL 2nd, Lawrence JD. Potentiation of warfarin by dong quai. *Pharmacotherapy.* 1999 Jul;19(7):870-6.

23. Heck AM, DeWitt BA, Lukes AL. Potential interactions between alternative therapies and warfarin. *Am J Health Syst Pharm.* 2000 Jul 1;57(13):1221-7. Review.

24. Belch JJ, Hill A. Evening primrose oil and borage oil in rheumatologic conditions. *Am J Clin Nutr.* 2000 Jan;71(1 Suppl):352S-6S.

25. Kremer JM. n-3 Fatty acid supplements in rheumatoid arthritis. *Am J Clin Nutr.* 2000;71(Suppl):349S-51S.

26. Pye JK, Mansel RE, Hughes LE. Clinical experience of drug treatments for mastalgia. *Lancet.* 1985;2:373-7.

27. Available at www.drweil.com. Accessed December 1, 2004.

28. Leonetti HB, Longo S, Anasti JN. Transdermal progesterone cream for vasomotor symptoms and postmenopausal bone loss. *Obstet Gynecol.* 1999;94:225-8.

29. Wren BG, Champion SM, Willetts K, et al. Transdermal progesterone and its effect on vasomotor symptoms, blood lipid levels, bone metabolic markers, moods, and quality of life for postmenopausal women. *Menopause.* 2003 Jan-Feb;10(1):13-8.

30. Available at www.acog.org/from_home/publications/press_releases/nr03-02-04.cfm. Accessed November 30, 2004.

CHAPTER 9

Depression

Christian R. Dolder

KEY POINTS

- Treating depression with a natural product has significant patient care concerns, including risk of suicide, management of poor response, and the possibility of confounding diseases—all of which may require the presence of a health care professional.
- The majority of clinical trials involving St. John's wort in mild-to-moderate depression are positive; however, the data supporting the use of St. John's wort in severe depression are not.
- St. John's wort is thought to work by a variety of mechanisms, including raising levels of serotonin, norepinephrine, and dopamine.
- St. John's wort can cause gastrointestinal, dermatological, and neurological side effects.
- St. John's wort is a broad-spectrum enzyme inducer and can lower the concentrations of many drugs. Care should be taken to check for drug interactions.
- St. John's wort can raise serotonin levels, which can become dangerous in a patient using other serotonergic agents.
- S-Adenosyl-L-methionine (SAMe) is an important endogenous compound that is ubiquitous in the human body.
- SAMe can raise serotonin and dopamine levels in the CNS, which may contribute to antidepressant activity.
- The majority of the limited study data available supports the use of SAMe in patients with depression.
- SAMe product quality can vary considerably.
- SAMe is well tolerated by most patients and has little significant drug interaction concerns.
- Inositol is being used for depression; however, study data are insufficient to recommend this use.
- Omega-3 fatty acids, including eicosapentaenoic acid (EPA) and docosahexaenoic acid (DHA), may be useful for depression.
- Fish oils are the most common source of omega-3 fatty acids.
- Side effects from fish oils are dose related; higher doses (> 3 g/day) cause a higher incidence of gastrointestinal side effects and increase bleeding risk.

Product	Dosage (Usual)	Effect	Safety Concerns
St. John's wort	300 mg three times daily	Majority of evidence supports efficacy for mild-to-moderate depression in adults; efficacy not present for severe depression; more studies needed in children	• Minor, most common side effects include gastrointestinal disturbances, skin reactions, sedation, restlessness or anxiety, dizziness, headache, and dry mouth; photosensitivity; and sexual side effects • Many drug interactions—see text • Avoid use in pregnancy and lactation
SAMe	400 to 800 mg/day for mild-to-moderate depression; 800 to 1600 mg/day for severe depression; also administered by injection	May be effective for depression; better trial data are required for conclusive statement on efficacy	• Minor gastrointestinal side effects • Significant variations in quality • Caution with other antidepressants; avoid in bipolar disorder • Insufficient pregnancy data to recommend use in early pregnancy; may be safe in third trimester • Insufficient lactation data to recommend use
Inositol	12 g/day	Inconclusive preliminary data	• Minor nausea, tiredness, headache, and dizziness • Insufficient pregnancy and lactation data to recommend use
Fish oils	Minimum 1 g/day	May improve symptoms of depression	• Nausea and loose stools at higher doses • Halitosis and belching • Can increase bleeding risk at higher doses (> 3 g/day)—use caution with concurrent use of antiplatelet/anticoagulant drugs and natural products that increase bleeding risk; monitor carefully for signs and symptoms of bleeding

In the United States, approximately 10% of people suffer from major depression at any one time, and 20% to 25% suffer an episode of major depression at some point during their lifetime.[1] Many patients, including children and elderly, are often untreated. This is tragic considering that this is a disorder with a variety of therapeutic options, including natural products.

The use of natural products in the treatment of depression is not without concern. Although natural products offer the possibility of increasing the number of people being treated for depression, such products also create the possibility of unsupervised medical care. For instance, adequate follow-up to monitor for suicidal ideations and therapeutic efficacy is crucial. Furthermore, different types of depression may respond to particular types of therapy; a competent clinician will be able to classify the depression type according to accepted guidelines and recommend a suitable treatment.

Most studies presented here compare the natural product to tricyclic antidepressants. These agents were widely used for many years as the drugs of choice for depression but fell out of favor with the approval of fluoxetine (Prozac®) in 1987. The tricyclic antidepressants are considered to be similar to the newer agents in efficacy.

PATIENT CASE

CLAUDIA

Claudia is a 34-year-old Caucasian female who complains of feeling depressed with symptoms of decreased sleep, reduced energy, muscle soreness, periods of tearfulness, and weight loss during the last month. She denies suicidal ideations. Claudia has a history of major depression. Her first episode was at the age of 25 with subsequent episodes at the ages of 27 and 32.

Claudia reports that her last depressive episode "was bad enough to make me go to the hospital." She reports having been prescribed a number of antidepressants in the past (paroxetine, sertraline, citalopram, and bupropion). While she agrees that these antidepressants were beneficial, she states that the side effects were "horrible" (i.e., sexual dysfunction, diarrhea, insomnia, sweating, and nervousness).

Claudia inquires about the use of an over-the-counter antidepressant. She realizes that she needs treatment and wants her depression to improve. At the same time, she would like you to recommend a more "natural medication" with fewer side effects.

ST. JOHN'S WORT

Documentation regarding the use of *Hypericum perforatum* (St. John's wort) extends over 2000 years to the time of the ancient Greeks. Currently, St. John's wort is widely used, especially in Europe, for the treatment of depression. St John's wort appears to produce its effects through a variety of biologically active constituents including hyperforin, adhyperforin, and hypericin. These active components produce their effects via a combination of mechanisms that are similar to, yet different from, common antidepressants.

St. John's wort

Early investigations reported that St. John's wort inhibited monoamine oxidase (MAO) types A and B; however, other investigators concluded that the MAO inhibition of St. John's wort was not potent enough to cause antidepressant effects.[2-5] A number of investigators have subsequently proposed that St. John's wort produces antidepressant effects through the inhibition of serotonin, norepinephrine, and dopamine synaptic reuptake.[2,6-8] Thus, it has been suggested that hypericum is somewhat similar to tricyclic antidepressants. The likelihood that St. John's wort is merely an herbal tricyclic antidepressant is unlikely due to a variety of additional mechanistic possibilities linked to hypericum. For instance, St. John's wort has also been shown to act on gamma-amino-butyric acid (GABA) and glutamate.

A number of controlled trials, meta-analyses, and reviews support the efficacy of St. John's wort in mild to moderate depression.[9-12] Taken together, these studies have reported the superiority of St. John's wort compared to placebo and its similarity compared to antidepressants (primarily tricyclic antidepressants). While there is a fair amount of supportive evidence, much of it is seriously flawed. Such problems include the use of nonstandardized diagnostic procedures, unvalidated symptom rating scales, and studies of short duration. Some investigations have compared St. John's wort to a standard antidepressant with only a small number of subjects and without a placebo group. This methodology could lead to equivalence between St. John's wort and its antidepressant comparator being erroneously reported.[13]

Unlike the findings from many investigations of hypericum's use in mild-to-moderate depression, several recent studies have created uncertainty about the use of this product in severe depression. Shelton and colleagues conducted an 8-week, randomized, double-blind, placebo-controlled trial of hypericum (900 to 1200 mg/day) in 200 outpatients with major depression.[13] At study completion, both the placebo and St. John's wort group had significant symptomatic improvement based on standardized symptom rating scales; however, no difference in improvement was found between the two groups.

The efficacy and safety of hypericum in major depressive disorder were tested in another 8-week, randomized, double-blind, placebo-controlled trial involving more than 300 outpatients.[14] The investigation also included a sertraline group. Patients received 900 to 1500 mg/day of hypericum, 50 to 100 mg/day of sertraline, or placebo. At 8 weeks, the two main outcome measures (depression rating scale and global impression scale) were not different between patients that had received St. John's wort or placebo. There was also no significant difference in efficacy when comparing the sertraline group to the placebo group.[13] Severe depression is difficult to treat, even with well-accepted therapies.

Safety concerns regarding the use of selective serotonin reuptake inhibitors in children and adolescents have sparked interest in the use of other antidepressants for this patient population. Recently, the results from an 8-week, open-label pilot study of St. John's wort in children and adolescents were reported. While this investigation did not have a control group, 25 of the 33 youths enrolled in the study met response criteria at the end of the trial. Twelve subjects reported side effects, which the investigators described as transient and mild.[15] More studies are needed regarding the use of St. John's wort in moderate-to-severe depression and in depressed children and adolescents.

A range of active ingredient concentrations and dosages have been used in trials of St. John's wort. Generally, most investigators have used St. John's wort extract containing 0.3% hypericin and a dose of 300 mg three times per day. Some studies have used doses up to 1200 or 1500 mg/day. Smaller doses (i.e., 300 to 600 mg/day) have been used in anxiety disorders and children. Similar to standard antidepressant medications, St. John's wort should not be abruptly discontinued due to the potential for withdrawal-related side effects.[16]

SAFETY CONSIDERATIONS/DRUG INTERACTIONS

The most common side effects reported in trials of St. John's wort include gastrointestinal disturbances, skin reactions, sedation, restlessness or anxiety, dizziness, headache, and dry mouth.[1] Dermatologic reactions have included rash, itching, pruritus, and photosensitization. Neurologically, cases of paresthesia and neuropathy have been documented. St. John's wort, like conventional antidepressants, has been reported to cause anxiety, mania, and serotonin syndrome. In addition, sexual side effects associated with St. John's wort are not uncommon. One 8-week trial of St. John's wort (900 to 1500 mg/day) found that 25% of patients receiving hypericum reported anorgasmia compared to 32% of patients taking sertraline.[14]

Several investigations and meta-analyses have concluded that while the above mentioned side effects do occur with the use of St. John's wort, many of these side effects occur at rates similar to or less than conventional antidepressants.[2,9,10,17,18] Nonetheless, St. John's wort is not for everyone. For instance, St. John's wort should be used with caution, if at all, in pregnancy and lactation. St. John's wort might have teratogenic effects and may cause colic, drowsiness, and lethargy in nursing infants.[19,20]

A number of significant drug-drug interactions, potential and actual, exist because of the ability of St. John's wort to induce the cytochrome CYP 450 liver enzyme system. This induction includes CYP 1A2, CYP 2C9, and especially the CYP 3A4 isoenzyme. St. John's wort was reported to increase CYP 3A4 activity by 98%. The increase in CYP 450 activity results in decreased therapeutic concentrations of concomitant medications that are also metabolized by the CYP 450 system.

St. John's wort has been shown to produce clinically significant decreases in the serum levels of medications including cyclosporine, oral contraceptives, simvastatin, protease inhibitors, tricyclic antidepressants, and warfarin. St. John's wort has also been reported to induce the intestinal P-glycoprotein drug transporter. This effect is thought to lead to St. John's ability to reduce digoxin drug levels. Interactions between St. John's wort and herbal medications are also possible. For instance, St. John's wort may decrease the serum level of red yeast rice and foxglove.[2]

Due to the serotonergic effects of St. John's wort, the combination of this herb with other medications having effects on serotonin must be used with caution. St. John's wort used in conjunction with selective serotonin reuptake inhibitors such as fluoxetine and paroxetine may lead to serotonin syndrome, a potentially life-threatening reaction. Theoretically, the combination of St. John's wort with a MAO inhibitor could lead to MAO inhibitor toxicity.[2]

S-ADENOSYL-L-METHIONINE (SAME)

SAMe is a supplement with a myriad of potential uses and associated mechanisms of actions. SAMe is synthesized by all living cells from the essential amino acid methionine and cannot be supplemented by diet. SAMe is a popular product that is being promoted as a treatment for such conditions as arthritis, liver disorders, migraine headaches, fibromyalgia, and depression. As an important source of methyl groups, SAMe is necessary in many transmethylation reactions responsible for the synthesis of a variety of molecules, including neurotransmitters such as serotonin and L-dopamine.[21]

Similar to standard antidepressant medications, the mechanism of action of SAMe is unclear. It is thought that SAMe's antidepressant effects are derived from its ability to promote high concentrations of serotonin and dopamine in the central nervous system. This might be accomplished by actions on the metabolism and reuptake of neurotransmitters. SAMe's ability to increase cell membrane fluidity may also enhance the activity of neurotransmitters by aiding the movement and binding of the transmitters.[21]

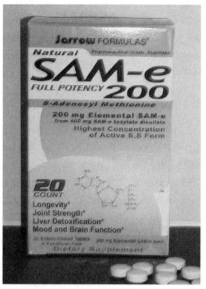

S-Adenosyl-L-methionine (SAMe)

Clinical trials examining the antidepressant effects of SAMe in patients with depression have been published. Most of these trials reported favorable results, with SAMe having antidepressant superiority compared to placebo and to tricyclic antidepressants. Short study durations (i.e., 4 weeks or less) of many of the investigations seriously limit the ability to state that SAMe has equivalent antidepressant effects compared to standard antidepressant medications. Nonetheless, the published literature does indicate an antidepressant effect for SAMe.

For instance, a 4-week, randomized, double-blind, controlled trial was conducted in patients with major depression. The 293 subjects were randomized to receive intramuscular SAMe or oral imipramine. At the end of the study, the two groups did not differ significantly on two depression rating scales or a global impression scale.[22] Similar results were reported in a 6-week trial in which nearly 300 patients were randomized to oral SAMe or oral imipramine.[23]

The most recently published meta-analysis of SAMe's antidepressant effects included six placebo-controlled studies and seven trials involving tricyclic antidepressants. When compared to placebo, SAMe had a global effect size of 17% to 38%. The author also found significant heterogeneity among the placebo-controlled trials, making generalizations difficult and also demonstrating the differences in efficacy between studies. Similar rates of response were found between SAMe and tricyclic antidepressants.[24]

SAMe is available in a variety of salt forms (sulfate, tosylate, and butanedisulfonate) from a number of manufacturers, creating a situation in which the actual amount of SAMe available in a given product may vary. In general, these formulations have very low bioavailability and the stability of the tosylate form is questionable. In terms of antidepressant dose, 400 to 800 mg/day orally is recommended for mild-to-moderate depression and 800 to 1600 mg/day orally for moderate-to-severe depression. Additionally, intramuscular dosing is most commonly administered weekly in doses ranging from 200 to 400 mg.[25]

Safety Considerations/Drug Interactions

Clinical trials of SAMe have not reported many significant side effects. Possible side effects include flatulence, nausea, vomiting, diarrhea, constipation, dry mouth, headache, mild insomnia, anorexia, sweating, dizziness, and nervousness.[24] The most common complaints consist of mild gastrointestinal adverse effects such as heartburn and constipation.

Many authors have specifically concluded that the incidence of side effects was less with SAMe compared to standard antidepressant medications, although most antidepressant trials involved a tricyclic antidepressant as the comparator agent. Data regarding the use of SAMe during lactation are inadequate and only a limited amount of safety information regarding the use of SAMe in pregnancy is available, although the use of SAMe during the third trimester may be safe.[25]

The drug interactions linked to SAMe are primarily theoretical and potential in nature. SAMe should be used cautiously with other antidepressants. A case of serotonin syndrome has been described in a patient taking both SAMe and clomipramine. The mild MAO inhibitory action of SAMe also creates the potential for hypertensive crisis associated with the ingestion of tyramine-containing foods.[26] Hypothetically, SAMe could increase the effects of natural products that also raise serotonin levels. SAMe should be avoided in patients with bipolar disorder. Theoretically, SAMe might methylate levodopa and worsen existing Parkinson's disease.[25]

Inositol

Inositol is an essential component of cell membrane phospholipids. Inositol is a metabolic precursor in the phosphatidylinositol cycle, a cycle that serves as a second messenger system for a variety of neurotransmitters.[27,28] The neurotransmitters involved in the phosphatidylinositol cycle include acetylcholine, serotonin, and norepinephrine.[29]

The potential relationship between inositol, the phosphatidylinositol cycle, and psychiatric disease has created interest in potential therapeutic efficacy. Individuals with bipolar or unipolar depression have been reported to have reduced levels of cerebrospinal inositol.[30] Postmortem studies of patients with schizophrenia and of psychiatric patients who had committed suicide reported reduced inositol levels.[31,32] One study reported that inositol supplementation significantly raised cerebrospinal inositol levels.[33]

Despite the interest in inositol's antidepressant activity, relatively few methodologically sound trials have been published. A 4-week, double-blind, placebo-controlled trial of inositol treatment was conducted in patients with unipolar depression (n = 22) or bipolar depression (n = 6). All of the patients had failed previous antidepressants or were intolerant to antidepressant medications. Inositol-treated subjects received 12 g/day. Per protocol, no other psychoactive medications were permitted except for benzodiazepines. Eleven of the 39 patients that originally consented dropped out and these patients were not included in the data analysis. At the end of 4 weeks, the inositol-treated group experienced a significant reduction in depressive symptoms compared to the placebo group.

Side effects reported in the inositol group included nausea and flatulence. No changes in hematology or liver and kidney function were found.[34] The antidepressant efficacy of inositol was also tested in a pilot study of 24 patients with symptomatic bipolar depression concomitantly receiving lithium, valproate, or carbamazepine.[35] While most of the study outcomes were not significantly different between the inositol and placebo groups, the positive direction of the results and tolerability of inositol may warrant a larger, controlled study.

There is no consensus regarding the antidepressant dose and duration of inositol. A dose of 12 g/day has been used in the few trials conducted for depression. The antidepressant effect of inositol has been reported to occur following at least several weeks of treatment.

Safety Considerations/Drug Interactions

The few trials of inositol have reported it to be well tolerated. The most common side effects were nausea, tiredness, headache, and dizziness.[36] Pregnancy and lactation data are insufficient for inositol. There are no known drug–drug or drug–natural product interactions reported with inositol.

Omega-3 Fatty Acids (Fish Oils)

Omega-3 fatty acids are long-chain, polyunsaturated fatty acids found in plant and marine life. Common examples are docosahexaenoic acid (DHA), eicosapentaenoic acid (EPA), and linolenic acid.[37] It is EPA and DHA that are thought to be pharmacologically active. Omega-3 fatty acids from such sources as fish oils (e.g., salmon and trout) have anti-inflammatory and antithrombotic effects via the inhibition of arachidonic acid-synthesized thromboxane A2 and vasodilatory effects.[38-40]

Fish oils are used in the treatment of hypertension, hypertriglyceridemia, Crohn's disease, rheumatoid arthritis, asthma, and cancer prevention. The arachidonic acid cascade is also involved in second messenger processes that affect the uptake of neurotransmitters.[38] Lipids constitute a major component of the human brain mass and are essential for normal brain structure and function.[41-43] Specific fatty acids in the brain (arachidonic acid, DHA, and EPA) play significant roles in neuronal signal transduction processes. The balance of these fatty acids may be altered in depression.[44] The use of omega-3 fatty acids in depression has been sparked by several additional observations:[45-47]

- Effect of omega-3 fatty acids on cerebrospinal serotonin metabolites
- Omega-3 fatty acid composition differences in depression
- Omega-3 fatty acid level differences in the diet and red blood cells of depressed individuals

Furthermore, a cohort study found that the amount of fish (a source of omega-3 fatty acids) consumed per capita was inversely related to the prevalence of depression.[48]

Although the biologic plausibility of omega-3 fatty acids' antidepressant effect has caused considerable interest, clinical trial evidence to support this excitement is just beginning to emerge. Nemets and colleagues conducted a 4-week, double-blind, placebo-controlled trial of omega-3 fatty acids as an adjunct to standard antidepressant treatment in patients with symptomatic unipolar depression. Nineteen of the 20 participants were receiving antidepressant treatment at the time of the study and were depressed despite the presence of stable antidepressant medication. Patients receiving omega-3 fatty acids were found to have a significant reduction from baseline in a depression rating scale. Furthermore, six of the ten omega-3 fatty-acid-treated patients and only one of ten of the placebo-treated individuals were considered responders.[49]

In a 4-month, double-blind, placebo-controlled study, omega-3 fatty acids were compared with placebo in 30 patients with bipolar disorder. The groups were given the study drugs in addition to their usual treatment. Individuals that received omega-3 fatty acids took a significantly longer time to become symptomatic and to require a change in medications.[50] The beneficial effects of omega-3 fatty acids were largely derived from its effect on study patients with depression rather than mania.

Su and colleagues reported the findings from an 8-week, double-blind, placebo-controlled study of omega-3 fatty acids in 28 outpatients with major depressive disorder. All subjects were symptomatic and all participants except for two were taking antidepressant medication. Participants in the treatment group received approximately 2 g of EPA and 1 g of DHA. Compared to the placebo group, patients in the omega-3 fatty acid group experienced significant improvement in a depression rating scale from the fourth week after treatment.[38]

Fish oils are the most common commercial source of omega-3 fatty acids. The antidepressant dose of fish oils is currently debatable. One depression study used 1 g of EPA twice daily; another administered 2 g of EPA and 1 g of DHA daily. A dose-ranging placebo-controlled study of EPA was conducted in 70 patients with symptomatic major depression despite ongoing treatment with standard antidepressant medications. Of the three doses studied (1, 2, and 4 g/day), the most evidence of efficacy was found in the 1g/day group.[44] Therefore, a dose of at least 1 g of EPA/day seems appropriate.

SAFETY CONSIDERATIONS/DRUG INTERACTIONS

Side effects associated with omega-3 fatty acids (i.e., fish oil) seem to be dose related to some extent. Fish oil products are generally well tolerated when administered in doses less than roughly 3 g/day. Higher doses cause more gastrointestinal side effects such as nausea and loose stools. Other side effects include belching, halitosis, and heartburn.[51-53] Higher doses of fish oils are also associated with the inhibition of platelet aggregation which can cause bleeding and theoretically increase the risk of hemorrhagic stroke.[54,55] The combination of fish oils with natural products having anticoagulant or antiplatelet effects (e.g., feverfew, garlic, ginger, ginkgo, and ginseng) could theoretically increase the risk of bleeding. A similar concern exists with medications possessing anticoagulant or antiplatelet effects (e.g., aspirin, warfarin, and others). A blood pressure lowering effect might cause hypotension in some patients. Fish oil products can also reduce vitamin E levels.[56]

Omega-3 fatty acids have been proposed to have a lack of teratogenicity and antidepressant protective effects for a fetus with a high risk of developing depression.[57,58] A case study reported the antidepressant effects of fish oil in a pregnant woman with a history of recurrent depressive episodes.[57] Regardless of the above information, there is insufficient pregnancy and lactation data for omega-3 fatty acids when used in medicinal quantities.[59] The doses used for depression are higher than those that can be reasonably obtained from food products. Consequently, concerns regarding contaminant exposure from eating fish are not discussed in this chapter.

Patient Discussion

Claudia is clearly depressed and meets diagnostic criteria for a major depressive episode. Initiation of an antidepressant would be beneficial. The addition of psychotherapy would also be advisable. Although Claudia has a number of effective prescription antidepressant medications to choose from, she is one of many antidepressant users that complain of side effects.

Selecting an effective antidepressant that does not cause intolerable side effects is Claudia's greatest therapeutic concern. In terms of natural products, there are not enough data to support the use of inositol or SAMe in this patient. St. John's wort or fish oil may be effective, but these agents' lack of supportive data in major depression is a concern. Claudia's previous intolerability to three selective serotonin reuptake inhibitors directs therapy to an antidepressant with an alternative mechanism of action such as mirtazapine or potentially venlafaxine, although venlafaxine has considerable serotonergic properties. If Claudia presented with a milder case of depression and/or without a history of depression, a trial of St. John's wort may have been advisable.

Natural products that can be tried in patients with depression include SAMe, St. John's wort, and fish oils. St. John's wort is the best-studied agent with evidence to support its use in mild-to-moderate depression. Evidence for efficacy in severe depression is not present. St. John's wort has considerable drug interactions and notable side effects. SAMe is the next best-studied agent. Data for fish oils are accumulating but are not yet conclusive. Patients that are at risk for suicide or have recurrent, chronic depression should be under the care of a health care professional.

Self-Assessment

1. A patient presents who meets criteria for major depressive disorder. The patient is also hearing voices and mumbling to herself. She states that she "strongly" dislikes taking drugs because they "mess up my body." This patient is a reasonable candidate for a trial of St. John's wort.
    a. True
    b. False

2. A patient who self-treats depression may not understand important concerns with this disease and therapy. However, since depression is often untreated, many patients might benefit from a trial of a natural product. Lack of treatment can have serious consequences.
    a. True
    b. False

3. A patient has depression that is refractive to several standard agents, including fluoxetine, citalopram, and venlafaxine. The depression has been categorized as severe. Since the patient has tried several prescription agents, the clinician should recommend inositol.
   a. True
   b. False

4. St. John's wort can interact with the following prescription medications:
   a. Birth control pills
   b. Medications used for HIV/AIDS
   c. Transplant drugs
   d. All of the above

5. The majority of clinical trial data supports the use of St. John's wort in patients with mild-to-moderate depression.
   a. True
   b. False

6. A patient wishes to use St. John's wort. The clinician should recommend the following nonprescription product:
   a. Aspirin, taken 1 hour beforehand
   b. Sunscreen, used as needed
   c. Psyllium, taken daily for constipation
   d. None of the above is necessary

7. A patient who has depression and an eating disorder was prescribed sertraline, which she uses daily. On her own initiative, she began regular use of St. John's wort and pseudoephedrine. She presents with symptoms of anxiety, agitation, and stomach upset. She is most likely experiencing:
   a. Generalized anxiety disorder (GAD)
   b. A stomach infection
   c. Serotonin syndrome
   d. An allergic reaction to St. John's wort

8. Which of the following is TRUE concerning SAMe:
   a. Product quality is not a concern.
   b. SAMe is not available in an oral formulation.
   c. SAMe is generally well tolerated.
   d. SAMe has significant, dangerous drug interactions that often prohibit the use of this product.

9. Which of the following is TRUE concerning fish oils:
   a. Clinical trial evidence is emerging that may support the use of fish oils as treatment for depression.
   b. Fish oils, used for depression, should be consumed as food products and not taken as supplements.
   c. Fish oils have notable side effects at doses greater than 10 g/day.
   d. Fish oils decrease bleeding risk in patients using aspirin or warfarin.

10. Which of the following components of fish oils are thought to provide benefit in psychiatric disease and other illnesses?
    a. EPA
    b. DHA
    c. ALA
    d. a and b

Answers: 1-b; 2-a; 3-b; 4-d; 5-a; 6-b; 7-c; 8-c; 9-a; 10-d

REFERENCES

1. Cleveland Clinic Mental Health Guide. Available at www.clevelandclinic.org.
2. Hammerness P, Basch E, Ulbricht C, et al. St. John's wort: a systematic review of adverse effects and drug interactions for the consultation psychiatrist. *Psychosomatics.* 2003;44:271-82.
3. Suzuki O, Katsumata Y, Oya M, et al. Inhibition of monoamine oxidase by hypericin. *Planta Med.* 1984;50:272-4.
4. Stock S, Holzi J. Pharmacokinetics test of [14C]labeled pseudohypericin from Hypericin perforatum and kinetics of hypericin in man. *Planta Med.* 1991;57:A61-2.
5. Thiede HM, Walper A. Inhibition of MAO and COMT by hypericum extracts and hypericin. *J Geriatr Psychiatry Neurol.* 1994;7(Suppl 1):S54-6.
6. Müller WE, Rolli M, Schafer C, et al. Effects of hypericum extract (LI 160) in biochemical models of antidepressant activity. *Pharmacopsychiatry.* 1997;30(Suppl 2):102-7.
7. Chatterjee SS, Bhattacharya SK, Singer A, et al. Hyperforin inhibits synaptosomal uptake of neurotransmitters in vitro and shows antidepressant activity in vivo. *Pharmazie.* 1998;53:9.
8. Neary JT, Bu Y. Hypericum LI 160 inhibits uptake of serotonin and norepinephrine in astrocytes. *Brain Res.* 1999;816:358-63.
9. Linde K, Ramirez G, Mulrow CD, et al. St. John's wort for depression—an overview and meta-analysis of randomized clinical trials. *BMJ.* 1996;313:253-8.
10. Kim HL, Streltzer J, Goebert D. St. John's wort for depression—a meta-analysis of well-defined clinical trials. *J Nerv Ment Dis.* 1999;187:532-8.

11. Lecrubier Y, Clerc G, Didi R, et al. Efficacy of St. John's wort extract WS 5570 in major depression: a double-blind, placebo-controlled trial. *Am J Psychiatry.* 2002;159:1361-6.

12. Van Gurp G, Meterissian GB, Haiek LN, et al. St. John's wort or sertraline? Randomized controlled trial in primary care. *Can Fam Physician.* 2002;48:905-12.

13. Shelton RC, Keller MV, Gelenberg A, et al. Effectiveness of St. John's wort in major depression: a randomized controlled trial. *JAMA.* 2001;285:1978-86.

14. Hypericum Depression Trial Study Group. Effect of Hypericum perforatum (St. John's wort) in major depressive disorder. *JAMA.* 2002;287:1807-14.

15. Findling RL, McNamara NK, O'Riordan MA, et al. An open-label pilot study of St. John's wort in juvenile depression. *J Am Acad Adolesc Psychiatry.* 2003;42:908-14.

16. Natural Medicines Comprehensive Database. St. John's wort monograph. Available at www.naturaldatabase.com.

17. Vorbach EU, Arnoldt KH, Hubner WD. Efficacy and tolerability of St. John's wort extract LI 160 versus imipramine in patients with severe depressive episodes according to ICD-10. *Pharmacopsychiatry.* 1997;30(Suppl 2):81-5.

18. Schrader E. Equivalence of St. John's wort extract (Ze 117) and fluoxetine: a randomized, controlled study in mild-moderate depression. *Int Clin Psychopharmacol.* 2000;15:61-8.

19. Chan LY, Chiu PY, Lau TK. A study of hypericin-induced teratogenicity during organogenesis using a whole rat embryo culture model. *Fertil Steril.* 2001;76:1073-4.

20. Lee A, Minhas R, Ito S, et al. Safety of St. John's wort during breastfeeding. *Clin Pharmacol Ther.* 2000;67:130, Abstract PII-64.

21. Nguyen M, Gregan A. S-Adenosylmethionine and depression. *Aust Fam Physician.* 2002;31:339-43.

22. Pancheri P, Scapicchio P, Chiaie RD. A double-blind, randomized parallel-group, efficacy and safety study of intramuscular S-adenosyl-L-methionine 1,4-butanedisulphonate (SAMe) versus imipramine in patients with major depressive disorder. *Int J Neuropsychopharmacol.* 2002;5:287-94.

23. Delle Chiaie R, Pancheri P, Scapicchio P. Efficacy and tolerability of oral and intramuscular S-adenosyl-L-methionine 1,4-butanedisulfonate (SAMe) in the treatment of major depression: comparison with imipramine in 2 multicenter studies. *Am J Clin Nutr.* 2002;76:1172S-6S.

24. Bressa GM. S-Adenosyl-l-methionine (SAMe) as antidepressant: meta-analysis of clinical studies. *Acta Neurol Scand.* 1994(Suppl); 154:7-14.

25. Natural Medicines Comprehensive Database. SAMe monograph. Available at www.naturaldatabase.com.

26. Fetrow CW, Avila JR. Efficacy of the dietary supplement S-adenosyl-L-methionine. *Ann Pharmacother.* 2001;35:1414-25.

27. Natural Medicines Comprehensive Database. Inositol monograph. Available at www.naturaldatabase.com.

28. Levine J. Controlled trials of inositol in psychiatry. *Eur Neuropsychopharmacol.* 1997;7:147-55.

29. Baraban JM, Worley PF, Snyder SH. Second messenger systems and psychoactive drug action: focus on the phosphoinositide system and lithium. *Am J Psychiatry.* 1989;146:1251-60.

30. Barkai AI, Dunner DL, Gross HA, et al. Reduced myo-inositol levels in cerebrospinal fluid from patients with affective disorder. *Biol Psychiatry.* 1978;13:65-72.

31. Shimon H, Agam G, Belmaker RH, et al. Reduced frontal cortex inositol levels in postmortem brain of suicide victims and patients with bipolar disorder. *Am J Psychiatry.* 1997;154:1148-50.

32. Shimon H, Sobolev Y, Davidson M, et al. Inositol levels are decreased in postmortem brain of schizophrenic patients. *Biol Psychiatry.* 1998;44:428-32.

33. Levine J, Gonzalves M, Babur I, et al. Inositol 6 gm daily may be effective in depression but not in schizophrenia. *Hum Psychopharmacol.* 1993;8:49-53.

34. Levine J, Barak Y, Gonzalves M, et al. Double blind controlled trial of inositol treatment of depression. *Am J Psychiatry.* 1995;152:792-4.

35. Chengappa KNR, Levine J, Gershon S, et al. Inositol as an add-on treatment for bipolar depression. *Bipolar Disord.* 2000;2:47-55.

36. Palatnik A, Frolov K, Fux M, et al. Double-blind, controlled, crossover trial of inositol versus fluvoxamine for the treatment of panic disorder. *J Clin Psychopharmacol.* 2001;21:335-9.

37. Freeman MP. Omega-3 fatty acids in psychiatry: a review. *Ann Clin Psychiatry.* 2000;12:159-65.

38. Su KP, Huang SY, Chiu CC, et al. Omega-3 fatty acids in major depressive disorder. A preliminary double-blind, placebo-controlled trial. *Eur Neuropsychopharmacol.* 2003;13:267-71.

39. Connor WE. n-3 Fatty acids from fish and fish oil: panacea or nostrum? *Am J Clin Nutr.* 2001;74;415-6.

40. Calder PC. N-3 polyunsaturated fatty acids, inflammation and immunity: pouring oil on troubled waters or another fishy tale? *Nutr Res.* 2001;21:309-41.

41. Horrobin DF, Bennett CN. New gene targets related to schizophrenia and other psychiatric disorders: enzymes, binding proteins and transport proteins involved in phospholipids and fatty acid metabolism. *Prostaglandins Leukot Essent Fatty Acids.* 1999;60:141-67. Review.

42. Horrobin DF. The membrane phospholipid hypothesis as a biochemical basis for the neurodevelopmental concept of schizophrenia. *Schizophr Res.* 1998;30:193-208.

43. Bennett CN, Horrobin DF. Gene targets related to phospholipids and fatty acid metabolism in schizophrenia and other psychiatric disorders: an update. *Prostaglandins Leukot Essent Fatty Acids.* 2000;63:47-59.

44. Peet M, Horrobin DF. A dose-ranging study of the effects of ethyl-eicosapentaenoate in patients with ongoing depression despite apparently adequate treatment with standard drugs. *Arch Gen Psychiatry.* 2003;59:913-9.

45. Hibbeln JR, Umhau JC, Linnoila M, et al. A replication study of violent and nonviolent subjects: cerebrospinal fluid metabolites of serotonin and dopamine are predicted by plasma essential fatty acids. *Biol Psychiatry.* 1998;44:243-9.

46. Edwards R, Peet M, Shay J, et al. Omega-3 polyunsaturated fatty acid levels in the diet and in red blood cell membranes of depression patients. *J Affect Disord.* 1998;48:149-55.

47. Maes M, Christophe A, Delanghe J, et al. Lowered omega3 polyunsaturated fatty acids in serum phospholipids and cholesteryl esters of depressed patients. *Psychiatry Res.* 1999;85:275-91.

48. Peet M, Murphy B, Shay J, et al. Depletion of omega-3 fatty acid levels in red blood cell membranes of depressive patients. *Biol Psychiatry.* 1998;43:315-9.

49. Nemets B, Stahl Z, Belmaker RH. Addition of omega-3 fatty acid to maintenance medication treatment for recurrent unipolar depressive disorder. *Am J Psychiatry.* 2002;159:477-9.

50. Stoll AL, Severus WE, Freeman MP, et al. Omega3 fatty acids in bipolar disorder: a preliminary double-blind, placebo-controlled trial. *Arch Gen Psychiatry.* 1999;56:407-12.

51. Ernst E, Rand JI, Barnes J, et al. Adverse effects profile of the herbal antidepressant St. John's wort (Hypericum perforatum L.). *Eur J Clin Pharmacol.* 1998;54:589-94. Review.

52. Belluzzi A, Brignola C, Campieri M, et al. Effects of new fish oil derivative on fatty acid phospholipid-membrane pattern in a group of Crohn's disease patients. *Dig Dis Sci.* 1994;39:2589-94.

53. Kris-Etherton PM, Harris WS, Appel LJ, et al. Fish consumption, fish oil, omega-3 fatty acids, and cardiovascular disease. *Circulation.* 2002;106:2747-57.

54. U.S. Food and Drug Administration. Center for Food Safety and Applied Nutrition. Letter regarding dietary supplement health claim for omega-3 fatty acids and coronary heart disease. Available at http://vm.cfsan.fda.gov/~dms/ds-ltr11.html.

55. Pedersen HS, Mulvad G, Seidelin KN, et al. N-3 fatty acids as a risk factor for haemorrhagic stroke. *Lancet.* 1999;353:812-3.

56. Meydani SN, Dinarello CA. Influence of dietary fatty acids on cytokine production and its clinical implications. *Nutr Clin Pract.* 1993;8:65-72.

57. Chiu CC, Huang SY, Shen WW, et al. Omega-3 fatty acids for depression in pregnancy. *Am J Psychiatry.* 2003;160:385.

58. Olsen SF, Sorensen JD, Secher NJ, et al. Randomised controlled trial of effect of fish-oil supplementation on pregnancy duration. *Lancet.* 1992;339:1003-7.

59. Natural Medicines Comprehensive Database. Fish oil monograph. Available at www.naturaldatabase.com.

Chapter 10
Cholesterol Reduction

Karen Shapiro

Currently, the recommended cholesterol levels from the National Cholesterol Education Program (NCEP) guidelines are 200 mg/dL or less for total cholesterol, 40 or higher for HDL, the "good" cholesterol, and (optimally) less than 100 for LDL, the "bad" cholesterol.[1] LDL levels may need to be lower than 100 (e.g., < 70 mg/dL) in patients at high cardiovascular risk. LDL levels greater than 100 are acceptable if the risk factors for coronary heart disease are low. Known risk factors for coronary heart disease include hypertension, smoking, HDL cholesterol < 40, advanced age ($\geq$ 45 years for men, ($\geq$ 55 years for women), or a family history of early heart disease.

Interestingly, patients often seek out natural products because they are wary of using HMG-CoA reductase inhibitors, or "statins" because of a perceived risk of memory problems. The data regarding statin-induced memory loss

KEY POINTS

- Patients with existing cardiovascular disease or at a high cardiovascular disease risk should use natural products as adjunctive therapies, if at all, and not as replacements for prescription agents.
- Oats, psyllium, and soy can modestly lower cholesterol.
- Bulk-forming agents need to be taken with adequate fluids.
- Soy must be used in food formulations (i.e., not as supplements) for cholesterol-lowering benefits.
- Clinicians will need to use dose–response relationships to ascertain benefit from these products and how they can be used to help individual patients reach goals of therapy.
- Plant sterols, available in margarines, supplements, and juices, can lower LDL cholesterol 7% to 15% when taken in the appropriate formulation and dose.
- Policosanol is used in many countries as a safe and effective therapy for cholesterol reduction.
- Garlic has modest benefit in cholesterol reduction.
- Product variability with garlic can be significant.
- Fish oils can significantly lower triglycerides and have anti-inflammatory properties.
- The most common side effects with fish oils are fishy-breath and burping.
- Fish oils, at doses greater than 3 g/day, can have an antiplatelet effect.

Product	Dosage	Effect	Safety Concerns
Oats	No supplement form—use as food product, e.g., 1 cup of oatmeal daily	Decreases LDL ~6 mg/dL	• Take at different times than oral drugs
Psyllium	10 to 12 g/day, divided; 3 g fiber = 6 Metamucil® capsules or 3 g fiber = 1 Metamucil wafer	Decreases LDL 5% to 15%	• Take at different times than oral drugs • Take with adequate fluids • Do not use if history of gut impaction, obstruction, or severe constipation
Soy	25 g/day	Decreases LDL 10%	• Avoid if history of breast cancer
Plant sterols	1 to 8 g, divided two or three times daily; 8 oz. Minute Maid Heart Wise® orange juice = 1 g sterols: capsule dosage varies—typical daily dose 1.8 g, taken in divided doses	Spreads, taken three times daily will decrease LDL 7% to 10%; juice, taken twice daily, will decrease LDL 12%; supplements, taken three times daily, will decrease LDL 10% to 15%	• Possible interaction with ezetimibe
Policosanol	5 to 10 mg twice daily	Decreases LDL 11% to 27%; increases HDL 11% to 26%	• May increase bleeding risk
Garlic	One clove fresh garlic daily or 300 to 400 mg three times daily, in a supplement standardized to contain 1.3% alliin	Decreases total cholesterol 4% to 6%	• Can increase bleeding risk • Malodorous breath or body odor
Fish oils	1 to 4 g, or more, daily; a typical capsule contains 500 mg EPA and DHA	Decreases triglycerides 25% to 50%	• Nausea and loose stools at higher doses • Halitosis, belching • May increase bleeding risk at higher doses (> 3 g/day)

(and possible benefits of statins for dementia prevention) are conflicting at present.[2] As with all medications the risks versus benefits need to be considered for each patient.

Niacin and red yeast products are not discussed. The guidelines from the American Heart Association (AHA) recommend against using dietary supplement niacin as a substitute for prescription niacin. Red yeast rice products contain mevinolin, an HMG-CoA reductase inhibitor with the same chemical structure as lovastatin. Red yeast rice products are illegal in the United States but are available to consumers on the Internet. The content of these products is unreliable and they can contain contaminants. They should not be recommended.

Oats

Food products containing whole oats are permitted by the FDA to claim that they lower cholesterol.[3] Oats are rich in beta-glucan, a water-soluble fiber that is responsible for many of the beneficial effects.[4-6] Beta-glucan dissolves in the large intestine and forms a gel, which binds to bile acids. Bile acids, when bound to fibers, are excreted in the feces, rather than being reabsorbed into the bloodstream.

The AHA Dietary Guidelines state that for every gram consumed from soluble-fiber rich foods, including oats (and psyllium, which is discussed next), the LDL should decrease by an average of 2.2 mg/dL.[7] Oatmeal will not be eaten throughout the day and is not available as a supplement. Three-quarters of a cup of dry oatmeal contains 3 g of oats. One cup of Cheerios® has 3.5 g.

PATIENT CASES

Tom—a "healthy" patient with high cholesterol

Tom is a 41-year-old Caucasian male. Tom has no past medical history except for minor sports-related injuries. He has no known current disease states. He plays tennis and basketball three or four times weekly. He does not smoke and consumes alcohol infrequently. His father is deceased from prostate cancer at age 67. His mother is alive and is described as having "excellent" health. Tom has normal body weight and blood pressure. His cholesterol panel is (mg/dL): CH 238, LDL 184, TG 149, and HDL 55.

Tom asks if he can buy any nonprescription product at the pharmacy to help lower his cholesterol. His doctor told him that his total cholesterol should be below 200 and his LDL, or "bad" cholesterol, should (optimally) be less than 100. Tom reports that his doctor told him to start using

PATIENT CASES *continued*

the "TLC" program, which includes watching his diet, or he might need to take prescription medicine. Tom states that he eats relatively healthy. He does not like brown bread or other "grains" but does eat lots of fruits and vegetables. He eats burger-type meals several times weekly (lunch) but tries to eat healthier evening meals. His breakfast consists of a muffin or pastry and coffee. He rarely snacks.

WINNIE—A PATIENT WITH HIGH TRIGLYCERIDES

Winnie is a 26-year-old Asian female. She reports no health problems until her pregnancy when her doctor noted that her triglycerides were high. She has been given a prescription for atorvastatin (Lipitor®) and plans to have it filled. She reports that her father had a heart attack at age 50 and is now deceased. Her mother is 54 years old and had a stroke this past year. She is the oldest of four siblings, all described as in good health. She does not smoke or drink.

Winnie is not taking any other medications except a multivitamin and calcium supplement. She states that she uses an IUD for contraception and does not plan on having more children. Her child is now 1 year old and she has stopped breastfeeding. She has taken a fasting lipid panel twice in the last month. She states her doctor wanted it repeated. Both panels had similar values. The last panel had the following values (mg/dL): CH 284, LDL 156, TG 390, and HDL 50.

LISA—A PATIENT WITH DIABETES AND HIGH CHOLESTEROL

Lisa is a 44-year-old Hispanic female. Lisa has diabetes type 2, controlled with metformin, glimepiride, and pioglitazone. Her blood pressure averages 142/90 mmHg. She is not taking any medications for the hypertension. She is 5'2" and weighs 148 pounds. She eats primarily beans, rice, cheese, and meat. Her vegetable intake consists of two or three servings weekly. She states that she used to walk nightly but has been too tired since she started her new job. Her cholesterol panel is (mg/dL): CH 206, LDL 137, TG 166, and HDL 35. She has been to her family doctor today who told her that she needs to take blood pressure and cholesterol medication. She's already taking five medications and feels that this is enough. Her other medications include aspirin for cardioprotection and citalopram for depression.

Safety Considerations/Drug Interactions

Oats are safe and offer benefits beyond cholesterol lowering. They provide satiety, can help normalize glucose levels, and contain important nutrients.[8] Because they are high in fiber, there is a possibility that the oats could bind to oral agents taken concurrently. It is best to administer oral medications separately from any high-fiber product. Oats and any other fiber-rich foods should be added to the diet gradually. Adding too much fiber at once can cause gas, bloating, or diarrhea. Fluid intake should be adequate or constipation can occur. Fluids may need to be increased as fiber intake is increased.

Oats

Psyllium

Psyllium is rich in soluble fibers. The primary mechanism for cholesterol lowering is the same as with oats. Like oats, psyllium has data supporting its role in controlling glucose levels.[9] Adding 10 to 12 g of psyllium daily can lower LDL levels by 5% to 15%.[10,11] It is difficult for many people to take psyllium mixed in water due to the grainy, somewhat acidic taste; however, many can tolerate Metamucil® capsules (3 g of fiber in six capsules) or wafers (3 g of fiber in one wafer). Since these products work primarily by binding to fat in the gut, they should be spread out among meals. Psyllium will provide little additional benefit when consumed at the same time as oats. Psyllium is in the class of laxatives called "bulk-forming" and is often recommended as the first choice for chronic constipation caused, in most cases, by the typical white-flour, low-fiber diet.

Safety Considerations/Drug Interactions

Psyllium has similar safety considerations as oats. It should be separated from oral medications and requires adequate fluid intake. In rare cases patients can be hypersensitive to psyllium and suffer severe allergic reactions.[12]

Soy

Soy can reduce LDL approximately 10% if taken in doses of approximately 25 g/day.[13-15] The FDA allows foods containing soy protein, included in a diet low in saturated fat and cholesterol, to state that consumption of these food products can reduce the risk of coronary heart disease. This health claim is limited to foods containing intact soy protein and does not include isolated substances from soy protein, such as soy supplements.

Soy, unlike some other beans, contains all the amino acids essential to human nutrition. Soy protein products can replace animal-based foods, which also contain complete proteins but tend to contain more fat, especially saturated fat. It would be beneficial if more ways could be found to increase soy consumption in the United States, for both personal health and environmental reasons. Use of animal meat as the primary source of dietary protein creates tremendous environmental waste and causes an increase in antibiotic resistance. Yet while the public is largely aware that soy is good for them, only 15% eat a soy product once weekly.[16]

Soy is high in fiber; however, the mechanism of action is different than that for oats and psyllium. Beta-sitosterol and other phytosterols, or plant sterols, are present in soy and contribute to the LDL-lowering effect by inhibiting cholesterol absorption. Sterols are marketed for cholesterol reduction in other products that contain higher sterol concentrations, which are discussed later. Another mechanism of action for soy is thought to involve activation of the LDL receptor—which could cause an increase in LDL uptake and degradation.[17,18] Since the mechanism of action is not thought to be attributable to the fiber content, it is reasonable to combine soy with oats or psyllium.

Acceptable soy food options must contain at least 6.25 g of soy protein per serving and be low in overall fat content, saturated fat, and cholesterol. The accompanying table lists soy foods that meet these requirements and can be recommended.[19] Soy food products come in many flavors and formulations, including cereals, sausages, "hotdogs," and "burgers."

Soy Content in Common Food Products

Soy product	Qty	Grams soy protein/serving
Whole cooked soybeans	1/2 cup	10 to 14
Roasted soy nuts	1/2 cup	12 to 15
Soy nut butter	2 tbsp.	30
Soy flour (whole grain)	1/2 cup	30
Soy flour (defatted)	1/4 cup	30
Tempeh	4 oz.	17
Firm tofu	4 oz.	13
Soft or silken tofu	4 oz.	9
Regular plain soymilk	8 oz.	10

Safety Considerations/Drug Interactions

Soy, taken in food form, is safe for all age groups. Although long-term soy intake may be protective against breast cancer,[20] it may not be wise to recommend soy products to women with a history of breast cancer, due to the possibility of estrogen-agonist effects. The majority of breast tumors grow in the presence of estrogen.

Plant Sterols

Plant sterols represent a safe, reputable option for patients. The NCEP guidelines include a recommendation to add plant sterols to the diet as part of the "TLC" program. As mentioned before, beta-sitosterol and other sterols are present in soy. Beta-sitosterol is one of several plant sterols and is found in almost all plants. The average intake of plant sterols in Western diets is less than 200 mg/day, while in Japan and among vegetarians it is nearly 400 mg/day. Cholesterol is the main animal sterol. Plant sterols are structured like cholesterol and, when present in sufficient amounts, block the micellar absorption of cholesterol.

However, whether one intakes 200 or 400 mg/day from food sources, at these levels sterols have little effect on cholesterol levels. Newer products marketed for cholesterol reduction contain plant sterols as additives in amounts much higher than found in foods naturally. These products are available as margarines, juices, and supplements. Margarines include Take Control® and Benecol®. Orange juice products include Minute Maid Heart Wise® Premium Orange Juice.[21]

Another benefit to using orange juice is that the juice itself can help raise HDL levels—but this requires drinking lots of juice—and consequently consuming lots of fructose.[22,23] The spreads or juices can work great for patients who already use a margarine—like spread on foods or drink orange juice. Patients can replace the product they are using with one of these agents. The margarine products, if used three times daily in place of other fats, can reduce LDL cholesterol by approximately 14%. The juice was shown to lower LDL levels by about 12% among adults who drank two glasses daily. Benecol spreads come in Regular and Light, a lower calorie spread. The regular Benecol spread can be used for cooking, baking, and frying-but not the other spreads because they do not contain enough fats. The Benecol Web site has recipes with nutritional information.[24]

Older persons may get more benefit with these products. In one study, subjects ages 50 to 59 reduced their LDL by an average of 21 mg/dL versus an average 17 mg/dL decrease in subjects 40 to 49 years old.[25] The benefit is significant for all age groups if the products complement a diet low in saturated fats and cholesterol.

Supplements include Benecol SoftGels, Preventive Nutrition Heart Advance, and others. Plant sterols have been reviewed for product quality by ConsumerLab and this site should be referenced to help select a quality product.[26]

Safety Considerations/Drug Interactions

These products are safe and well tolerated. In some patients, especially with excessive use, indigestion-type problems may occur. There may be an interaction between plant sterols and the prescription drug ezetimibe but this is currently not known. This drug works by a mechanism similar to plant sterols. If there is an interaction, it is likely that one agent changes the efficacy of the other. Ezetimibe is approved for both high cholesterol and a rare genetic condition called sitosterolemia in which affected individuals absorb plant sterols. Patients with sitosterolemia need to avoid plant sterols in any formulation, including food products and cholesterol-lowering agents.

Policosanol

Policosanol is a mixture of waxy substances generally manufactured from sugarcane. Some available products are made from beeswax; however, there is reason for concern that these products may not be as effective. Numerous studies have shown that policosanol can lower LDL cholesterol 17% to 27% and increase HDL cholesterol from 11% to 26%.[27] Policosanol is thought to work by decreasing cholesterol synthesis and increasing LDL clearance.[23,28,29] This agent is approved as a treatment for high cholesterol in about two dozen countries.

Policosanol also appears to be helpful for intermittent claudication, a disease caused by hardening of the arteries. It inhibits abnormal platelet aggregation, protects against LDL oxidation, and suppresses arterial inflammatory factors.[30,31] The mechanism of action for cholesterol reduction is not completely clear but appears to involve inhibition of cholesterol synthesis and enhancement of LDL degradation.[25,31,32] A typical dose is 5 to 10 mg twice daily. (Average daily dose in clinical trials is 12 mg.)

Safety Considerations/Drug Interactions

Policosanol is used by millions daily and has an excellent safety record. It is free of significant adverse effects. There is no effect on liver enzymes or muscle toxicity. Since this agent is the most popular natural product for cholesterol lowering internationally, it is worth repeating that health benefits seen with statins may or may not be present with this or any of the other agents discussed. Mortality benefit is also not known. Policosanol can decrease platelet aggregation and should be used carefully or avoided in patients taking concurrent antiplatelet drugs (e.g., aspirin and clopidogrel), anticoagulants (e.g., warfarin), natural products that can increase bleeding risk (e.g., dong quai, feverfew, fish oils, garlic, ginger, ginseng, and vitamin E) or in patients with a history of bleeding.[33]

GARLIC

Well-designed studies have demonstrated that garlic can modestly lower cholesterol and blood pressure.[34-36] Garlic has antiviral properties. Andrew Weil, MD, a leader in natural and preventive medicine, claims that the best home remedy for colds is to eat several cloves of garlic at the first onset of symptoms.[37] (He also recommends chewing fresh parsley afterwards to minimize the odor.)

Considerations in using garlic for cholesterol reduction are twofold. First, the results that can be obtained are less than can be obtained with other agents. A meta-analysis of well-designed trials with quality garlic preparations found that garlic use decreased total cholesterol by 4% to 6%, which is similar to the effects of 6 months of dietary interventions.[38,39] In this study, the authors concluded that "...garlic is superior to placebo in reducing total cholesterol levels. However, the size of the effect is modest, and the robustness of the effect is debatable."

The second consideration has to do with the comment that "the robustness of the effect is debatable." This refers to the widely variable trial results regarding efficacy, resulting from a lack of standardization in commercial products. Allicin, the active ingredient in garlic, is strongest in the fresh herb and varies in commercial preparations. Most of the positive clinical trials have been performed with garlic yielding at least 3,600 to 5,400 mcg of allicin per day, or the amount from a bulb of fresh garlic.

For cholesterol reduction and other health benefits, fresh herb is preferable; however, the herb needs to be raw in order to retain the allicin content and raw herb can cause upset stomach and heartburn, especially in patients who are not used to consuming fresh garlic. Supplements can be used, but product quality should be expected to be highly variable. The ConsumerLab Web site or another reputable resource should be used to choose a suitable product. An average dose used in studies is 300 to 400 mg of a product standardized to contain 1.3% alliin, taken three times daily.

Commercial preparations often state aliin content. Aliin is converted by the enzyme alliinase to allicin. This conversion consequently depends on the amount of aliinase present in the supplement, which is also variable.[40] Most preparations that are standardized come in dosages of 300 to 400 mg per tablet, or higher in some enteric-coated formulations.

SAFETY CONSIDERATIONS/DRUG INTERACTIONS

The most common side effect from garlic is breath odor, which occurs commonly with both regular and "odorless" garlic preparations. Raw garlic can cause stomach upset and heartburn, particularly with excessive doses. Garlic can induce hepatic enzymes and may lower the concentration of substrate drugs. This primarily affects CYP 450 3A4 and the effect is variable among different garlic supplements.[41,42] Garlic can increase the risk of bleeding, possibly due to an antiplatelet effect.[43,44] Use caution in patients at an increased bleeding risk, as described earlier.

FISH OILS (OMEGA-3 FATTY ACIDS)

Sources of omega-3 fatty acids

Fish oil is a rich source of the omega-3 fatty acids, eicosapentaenoic acid (EPA) and docosahexaenoic acid (DHA). Omega-3 fatty acids are one of the two main classes of essential fatty acids (EFAs). A third omega-3, called alpha-linolenic acid (ALA), is found primarily in dark green leafy vegetables, flaxseed oil, and some vegetable oils. The body has enzymes that convert ALA to EPA; however, the conversion rate is low. All three are important to human health. Omega-6 fatty acids are the other main class of EFAs and come primarily from plants.

Omega-3 fatty acids have an important role in nutrition. They are significant structural components of the phospholipid cell membranes, particularly in the brain, retina, and other nerve tissue. They are essential for the formation of new tissue and are important for fetal and infant development. They can inhibit the synthesis of pro-inflammatory substances and are useful for conditions such as rheumatoid arthritis and psoriasis. They inhibit platelet aggregation, stabilize cardiac conductivity, lower triglyceride levels, reduce blood pressure modestly, and improve arterial function.

Much of the research into the potential therapeutic benefits of omega-3 fatty acids began with studies of the Inuit people who live near the Arctic. The traditional Inuit diet consists of a large amount of "fatty fish," including seals, whales, and walrus, yet the Inuit seldom suffer from heart disease. When the Inuit consume a typical Western style diet, the risk of heart disease is about the same as the average North American.

An appropriate balance between omega-3 and omega-6 fats in the diet is important for good health.[45-47] Both are components of cell membranes and compete with each other for incorporation into the lipid bilayer. Replacement of EPA and DHA with arachidonic acid (a derivative of omega-6 fatty acids) leads to a more inflammatory, thrombogenic state.[48] Ideally, the ratio of omega-3 to omega-6 fatty acids in the human body should be roughly equivalent. However, because the typical American diet is low in omega-3s and high in omega-6s, many people have 10 to 20 times more omega-6 fatty acids than omega-3 fatty acids in their systems. One might think you would need to substantially change this ratio in order to provide benefit, but it appears that omega-3 supplementation is beneficial even in the presence of high omega-6 consumption.[49]

Eating fish is a healthy way to intake beneficial omega-3's, yet seafood is consumed much less than red meat and poultry as a source of protein, and the seafood industry reports that consumption has fallen from 1987 to 2001.[50] Recent warnings regarding mercury intake from fish sources must be considered. Mercury is toxic to the developing brain and nervous system. According to the Centers for Disease Control and Prevention, 8% of U.S. women of childbearing age have levels of mercury in their blood that present developmental risks for their babies.[51] A 2004 joint FDA/EPA health advisory counsels that pregnant women should not eat shark, swordfish, king mackerel, or tilefish at all and should limit weekly consumption of other fish and shellfish to 12 ounces (two average meals) a week.[52] Salmon and canned light tuna, two of the most commonly eaten fish, are low in mercury; however, farmed raised salmon (which is less expensive than wild salmon) has seven times the PCBs and dioxins of wild salmon. PCBs and dioxins are harmful contaminants.[53]

EPA and DHA Content in Fish

Type of fish, 3 oz.	EPA + DHA, in grams
Tuna, fresh	0.28 to 1.51
Tuna, light, canned in water, drained	0.26
Tuna, white, canned in water, drained	0.73
Sardines	0.98 to 1.70
Atlantic salmon	1.28 to 2.15
Mackerel	0.4 to 1.85
Rainbow trout	1.15
Flounder or sole	0.49
Shrimp	0.32

Source: Adapted from Reference 57.

For people who do not eat enough fish and shellfish or wish to avoid contaminants, supplements made from salmon or other fish can fill the gap. Cod liver oil supplements are rich in omega-3s but should *not* be recommended due to the risk of vitamin A and D toxicity. Consumer Reports and ConsumerLab each tested a variety of fish oil supplements and found that they were safe as far as contaminants and decomposition were concerned.[54,55] None of the products was found to contain detectable levels of mercury.[54,55] The quantity of omega-3s can be problematic. The ConsumerLab study found that six of twenty products tested had lower levels of omega-3s than were listed on the label.

How much fish oils does one need to take? For protection (primary prevention) against coronary heat disease, the AHA recommends that everyone eat at least two 3-ounce servings of fish a week. How much this represents in DHA and EPA depends on the type of fish, where it was from, how it was raised, and packaging and cooking methods. The accompanying table represents approximate EPA and DHA content per serving. For patients with elevated triglycerides, 4 g/day can lower triglycerides 25% to 50%.[56]

Safety Considerations/Drug Interactions

A problem that patients might experience with fish oil supplements is fish-odor when burping or breathing. Nausea and loose stools can occur with higher doses. The capsules are best taken at night or with food to help with this problem; however, capsules are large and the dose may need to be split up. Fish oils have an antiplatelet effect at doses greater than 3 g/day. Use caution in patients at an increased bleeding risk, as described earlier.[33]

Patient Discussion

A "Healthy" Patient with High Cholesterol

Considering that Tom has zero risk factors at this time, his LDL should be less than 160. Lower than this would be preferable. By reaching LDL goals, the cholesterol levels should fall closer to the recommended total cholesterol of less than 200, but they will not reach this level unless his LDL drops significantly below 160. High triglycerides are considered an independent risk factor for coronary heart disease. Tom's triglycerides are at the high end of normal.

Tom's physician was correct in counseling him to begin "Therapeutic Lifestyle Changes," or TLC, since his LDL cholesterol is high. TLC focuses on diet, weight reduction, and increased physical activity. Tom does not seem eager to consume whole grain products in the form of bread or rice, but he might consider using other food products that can lower cholesterol. Oat-containing cereals, such as Cheerios or oatmeal, would make a healthy substitution of the muffin or pastry he currently consumes for breakfast. Prepared baked goods are typically high in saturated fats, the worst types of fats for the heart, and the TLC program includes reducing saturated fat intake.

To reach an LDL below 160, Tom needs a reduction of at least 24 mg/dL, or 13%. To reach an LDL of 115 (which would bring his total cholesterol within normal values), he would need a reduction of at least 69 mg/dL, or 38%. When prescription medication is used to reduce a patient's cholesterol, the dose necessary to reach the desired effect is approximated. Approximating effect from natural products is also necessary. If one product can give an LDL reduction of 5% to 15%, but the patient needs a 25% reduction, then it is likely that combination agents (e.g., similar to prescription product combinations) will be necessary.

Tom can use combination natural products to reach a safer LDL level—but to get to an optimal level will be difficult without the use of prescription products. Patients also need to understand that the benefit of these products work only as long as the product is being used—unless the change is also due to concurrent dietary and other lifestyle changes. For example, plant sterols impair cholesterol absorption, causing blood cholesterol levels to decrease. As soon as a person stops using plant sterols, the body immediately starts to absorb more cholesterol and the blood cholesterol rises. Another consideration with the use of natural products is that other benefits seen with prescription drugs, such as plaque stabilization from statins, may not be present.

A PATIENT WITH HIGH TRIGLYCERIDES

Winnie's physician has prescribed atorvastatin, which is an acceptable choice unless she becomes pregnant. She is using an IUD and runs minimal risk of accidental pregnancy with this form of contraception. She may carry a genetic risk for high triglycerides and, consequently, premature coronary heart disease. Fish oil is an appropriate option to lower triglycerides but may not be enough for this patient. She may be able to reduce her triglycerides to around 200 mg/dL with fish oil alone, but this will remain at an unhealthy level. She will need to use the prescription agent and can use fish oils concurrently. It is best not to start both agents at the same time so that the change from one agent can be measured. Fish oils have anti-inflammatory properties and offer benefits beyond lowering triglycerides, including a decrease in the risk of cardiovascular mortality.[58]

If she does attempt to use fish oils alone, she should have her lipid panel repeated 3 months after using at least 4 g/day.

A PATIENT WITH DIABETES AND HIGH CHOLESTEROL

Lisa has diabetes with elevated cholesterol and hypertension.[59] She is at very high risk for coronary heart disease. Her LDL goal is <70 mg/dL.[60] She should be on a statin and her lipid values should be normalized as quickly as possible. She should be counseled on the need for the additional medications. Her wish to use natural products can be channeled into the need to eat healthier food choices and increase her level of physical activity, both natural ways to help control her conditions. She can be counseled to replace dietary fats with plant sterol spreads and increase her consumption of high-fiber foods, including oats.

SELF-ASSESSMENT

1. Which is TRUE concerning garlic:
   a. Garlic can modestly lower blood pressure and cholesterol.
   b. There is significant variation in product quality.
   c. Garlic has an antiplatelet effect.
   d. All of the above.

2. Which of the following food products is permitted by the FDA to claim on package labeling that it lowers cholesterol?
   a. White rice
   b. Oats
   c. Peanut butter
   d. Salmon

3. Which is TRUE concerning plant sterols:
   a. All orange juice products contain plant sterols.
   b. Plant sterols have an antiplatelet effect.
   c. Plant sterol products include juices, spreads, and supplements.
   d. It is not necessary to consume a low-fat diet when using plant sterol products.

4. Which is TRUE concerning fish oils:
   a. 4 g/day can lower triglycerides 25% to 50%.
   b. 4 g/day is a low dose and will not affect platelets.
   c. Fish oil supplements are usually contaminated with mercury.
   d. Fish oil supplements increase inflammation.

5. A female of childbearing age wishes to consume daily omega-3 fatty acids to help with an inflammatory condition. Which of the following represents her best treatment option:
   a. Salmon
   b. Cod liver oil
   c. Fish oil supplements made from salmon or other fish
   d. Tuna

6. Which is TRUE concerning soy:
   a. Soy supplements are recommended for lowering cholesterol.
   b. Soy food products are low in fiber.
   c. Soy products have many health benefits.
   d. A cup of soy milk has more grams of protein than a cup of soy beans.

7. What dosage forms are available for a person who wishes to use psyllium to help lower cholesterol?
   a. Powder
   b. Wafers
   c. Supplements
   d. All of the above

8. A female patient has diabetes. Her LDL is 137 mg/dL. She has tried TLC for 6 months with little effect. Her physician has given her a prescription for pravastatin. She is interested in using a natural product instead. You should counsel her that the most appropriate treatment option is:
   a. A prescription HMG Co-A reductase inhibitor, such as pravastatin
   b. Combination natural products
   c. Fish oils, at high doses
   d. Plant sterol supplements

9. Which of the following natural products would be expected to lower LDL to the greatest extent?
   a. Garlic
   b. Fish oils
   c. Plant sterols
   d. Oats

10. Soy supplements should be avoided in patients with a history of:
    a. Breast cancer
    b. Hypertension
    c. High cholesterol
    d. Eczema

Answers: 1-d; 2-b; 3-c; 4-a; 5-c; 6-c; 7-d; 8-a; 9-c; 10-a

REFERENCES

1. Grundy SM, Becker D, Clark LT, et al. Executive summary of the third report of the National Cholesterol Education Program (NCEP) Expert Panel on Detection, Evaluation, and Treatment of High Blood Cholesterol in Adults (Adult Treatment Panel III). JAMA. 2001;285:2486-97.

2. Rodriguez EG, Dodge HH, Birzescu MA, et al. Use of lipid-lowering drugs in older adults with and without dementia: a community-based epidemiological study. J Am Geriatr Soc. 2002;50:1852-6.

3. Available at www.fda.gov/bbs/topics/ANSWERS/ ANS00782.html Accessed July 12, 2004.

4. Braaten JT, Wood PJ, Scott FW, et al. Oat beta-glucan reduces blood cholesterol concentration in hypercholesterolemic subjects. Eur J Clin Nutr. 1994;48:465-74.

5. Brown L, Rosner B, Willett WW, et al. Cholesterol-lowering effects of dietary fiber: a meta-analysis. Am J Clin Nutr. 1999;69:30-42.

6. Ripsen CM, Keenan JM, Jacobs DR, et al. Oat products and lipid lowering. A meta-analysis. JAMA. 1992;267:3317-25.

7. Krauss RM, Eckel RH, Howard B, et al. AHA Dietary Guidelines: Revision 2000: A Statement for Healthcare Professionals from the Nutrition Committee of the American Heart Association. Circulation. 2000;102:2284-99.

8. Hallfrisch J, Scholfield DJ, Behall KM. Diets containing soluble oat extracts improve glucose and insulin responses of moderately hypercholesterolemic men and women. Am J Clin Nutr. 1995;61:379-84.

9. Anderson JW, Allgood LD, Turner J, et al. Effects of psyllium on glucose and serum lipid responses in men with type 2 diabetes and hypercholesterolemia. Am J Clin Nutr. 1999;70:466-73.

10. Schectman G, Hiatt J, Hartz A. Evaluation of the effectiveness of lipid-lowering therapy (bile acid sequestrants, niacin, psyllium and lovastatin) for treating hypercholesterolemia in veterans. Am J Cardiol. 1993;71:759-65.

11. Sprecher DL, Harris BV, Goldberg AC. Efficacy of psyllium in reducing serum cholesterol levels in hypercholesterolemic patients on high- or low-fat diets. Ann Intern Med. 1993;119:545-54.

12. Lantner RR, Espiritu BR, Zumerchik P, et al. Anaphylaxis following ingestion of a psyllium-containing cereal. JAMA. 1990;264:2534-6.

13. Matvienko OA, Lewis DS, Swanson M, et al. A single daily dose of soybean phytosterols in ground beef decreases serum total cholesterol and LDL cholesterol in young, mildly hypercholesterolemic men. Am J Clin Nutr. 2002;76:57-64.

14. Washburn S, Burke GL, Morgan T, et al. Effect of soy protein supplementation on serum lipoproteins, blood pressure, and menopausal symptoms in peri-menopausal women. Menopause. 1999;6:7-13.

15. Bakhit RM, Klein BP, Essex-Sorlie D, et al. Intake of 25 g of soybean protein with or without soybean fiber alters plasma lipids in men with elevated cholesterol concentrations. *J Nutr.* 1994;124:213-22.

16. Henkel J. Soy: health claims for soy protein, questions about other components. Available at www.fda.gov/fdac/features/2000/300_soy.html. Accessed July 14, 2004.

17. FDA Approves New Health Claim for Soy Protein and Coronary Heart Disease, Food and Drug Administration (FDA), U.S. Department of Health and Human Services, October 20, 1999. Available at www.fda.gov/bbs/topics/ANSWERS/ANS00980.html. Accessed July 13, 2004.

18. Sirtori CR, Lovati MR. Soy proteins and cardiovascular disease. *Curr Atheroscler Rep.* 2001 Jan;3(1):47-53.

19. Wart PJ. Soy: Not Just for Vegetarians Anymore. Vanderbilt University HealthPlus. Available at http://vanderbiltowc.wellsource.com/dh/Content.asp?ID=251. Accessed July 13, 2004.

20. Murkies A, Dalais FS, Briganti EM, et al. Phytoestrogens and breast cancer in postmenopausal women: a case control study. *Menopause.* 2000;7:289-96.

21. Devaraj S, Jialal I, Vega-Lopez S. Plant sterol-fortified orange juice effectively lowers cholesterol levels in mildly hypercholesterolemic healthy individuals. *Arterioscler Thromb Vasc Biol.* 2004 Mar;24(3):E25-8. Epub Feb. 5, 2004.

22. Miettinen TA, Gylling H. Plant stanol and sterol esters in prevention of cardiovascular diseases. *Ann Med.* 2004;36(2):126-34.

23. Kurowska EM, Spence JD, Jordan J, et al. HDL-cholesterol-raising effect of orange juice in subjects with hypercholesterolemia. *Am J Clin Nutr.* 2000 Nov;72(5):1095-100.

24. Benecol Recipe Library. Available at www.benecol.com/recipes/index.jhtml. Accessed July 10, 2004.

25. Law M. Plant sterol and stanol margarines and health. *BMJ.* 2000;320:861-64.

26. Product Review: Cholesterol-Lowering Supplements (Guggulsterones, Policosanol, Sterols). Available at www.consumerlab.com/results/cholest.asp. Accessed July 6, 2004.

27. Varady KA, Wang Y, Jones PJ. Role of policosanols in the prevention and treatment of cardiovascular disease. *Nutr Rev.* 2003 Nov;61:11:376-83.

28. Pepping J. Policosanol. *Am J Health Syst Pharm.* 2003;60(11):1112-5.

29. McCarty MF. Policosanol safely down-regulates HMG-CoA reductase-potential as a component of the Esselstyn regimen. *Med Hypotheses.* 2002 Sep;59(3):268-79.

30. Castano G, Mas R, Fernandez JC, et al. Effects of policosanol 20 versus 40 mg/day in the treatment of patients with type II hypercholesterolemia: a 6-month double-blind study. *Int J Clin Pharmacol Res.* 2001;21(1):43-57.

31. Gouni-Berthold I, Berthold HK. Policosanol: clinical pharmacology and therapeutic significance of a new lipid-lowering agent. *Am Heart J.* 2002;143:356-65.

32. Menendez R, Amor AM, Rodeiro I, et al. Policosanol modulates HMG-CoA reductase activity in cultured fibroblasts. *Arch Med Res.* 2001 Jan-Feb;32:1:8-12.

33. Arruzazabala ML, Valdes S, Mas R, et al. Comparative study of policosanol, aspirin and the combination therapy policosanol-aspirin on platelet aggregation in healthy volunteers. *Pharmacol Res.* 1997;36:293-7.

34. McMahon FG, Vargas R. Can garlic lower blood pressure? A pilot study. *Pharmacotherapy.* 1993 Jul-Aug;13(4):406-7.

35. Auer W, Eiber A, Hertkorn E, et al. Hypertension and hyperlipidaemia: garlic helps in mild cases. *Br J Clin Pract Symp Suppl.* 1990;69:3-6.

36. Silagy CA, Neil HA. A meta-analysis of the effect of garlic on blood pressure. *J Hypertens.* 1994 Apr;12(4):463-8.

37. Weil A. *Natural Health, Natural Medicine.* New York:Houghton Mifflin Co; 1998:241.

38. Stevinson C, Pittler MH, Ernst E. Garlic for treating hypercholesterolemia: a meta-analysis of randomized clinical trials. *Ann Intern Med.* 2000;133:420-9.

39. Tang JL, Armitage JM, Lancaster T, et al. Systematic review of dietary intervention trials to lower blood total cholesterol in free-living subjects. *BMJ.* 1998;316:1213-20.

40. ConsumerLab Product Review: Garlic supplements. Available at www.consumerlab.com/results/garlic.asp. Accessed July 18, 2004.

41. Gurley BJ, Gardner SF, Hubbard MA, et al. Cytochrome P450 phenotypic ratios for predicting herb-drug interactions in humans. *Clin Pharmacol Ther.* 2002 Sep;72(3):276-87.

42. Sparreboom A, Cox MC, Acharya MR, et al. Herbal remedies in the United States: potential adverse interactions with anticancer agents. *J Clin Oncol.* 2004 Jun 15;22(12):2489-503. Review.

43. Sunter WH. Warfarin and garlic. *Pharm J.* 1991;246:722.

44. Golovchenko I, Yang CH, Goalstone ML, et al. Garlic extract methylallyl thiosulfinate blocks insulin potentiation of platelet-derived growth factor-stimulated migration of vascular smooth muscle cells. *Metabolism.* 2003 Feb;52(2):254-9.

45. Holub BJ. Fish oils and cardiovascular disease. *CMAJ.* 1989;141:1063.

46. Kris-Etherton PM, Taylor DS, Yu-Poth S, et al. Polyunsaturated fatty acids in the food chain in the United States. *Am J Clin Nutr.* 2000;71(1Suppl):179S-88S.

47. Schmidt EB, Skou HA, Christensen JH, et al. n-3 Fatty acids from fish and coronary artery disease: implications for public health. *Public Health Nutr.* 2000;3:1:91-8.

48. Harper CR, Jacobson TA. Beyond the Mediterranean diet: the role of omega-3 Fatty acids in the prevention of coronary heart disease. *Prev Cardiol.* 2003;6:136-46.

49. Hwang DH, Chanmugam PS, Ryan DH, et al. Does vegetable oil attenuate the beneficial effects of fish oil in reducing risk factors for cardiovascular disease? *Am J Clin Nutr.* 1997;66:89-96.

50. Brody JE. Tip the Scale in Favor of Fish: The Healthful Benefits Await. *NY Times.* July 29, 2003.

51. Blood and Hair Mercury Levels in Young Children and Women of Childbearing Age—United States, 1999. MMWR. March 02, 2001 / 50(08);140-3. Available at www.cdc.gov/mmwr/preview/mmwrhtml/ mm5008a2.htm. Accessed July 9, 2004.

52. What You Need to Know about Mercury in Fish and Shellfish. FDA/EPA EPA-823-R-04-005. March 2004. Available at www.cfsan.fda.gov/~dms/admehg3.html. Accessed July 18, 2004.

53. Hites RA, Foran JA, Carpenter DO, et al. Global assessment of organic contaminants in farmed salmon. *Science.* 9 January 2004;303:226-9.

54. Omega-3 fatty acids. *Consumer Reports.* October 2003.

55. ConsumerLab Product Review: Omega-3 Fatty Acids (EPA and DHA) from Fish/Marine Oils. Available atwww.consumerlab. com/results/omega3.asp. Accessed July 18, 2004.

56. Din JN, Newby DE, Flapan AD. Omega-3 fatty acids and cardiovascular disease—fishing for a natural treatment. *BMJ.* 2004;328(7430):30-5.

57. Kris-Etherton PM, Harris WS, Appel LJ. American Heart Association Nutrition Committee. Fish consumption, fish oil, omega-3 fatty acids, and cardiovascular disease. *Circulation.* 2002;106(21):2747-57.

58. Studer M, Briel M, Leimenstoll B, et al. Effect of different antilipidemic agents and diets on mortality: a systematic review. *Arch Intern Med.* 2005;165:725-30.

59. Chobanian AV, Bakris GL, Black HR, et al. The Seventh Report of the Joint National Committee on Prevention, Detection, Evaluation, and Treatment of High Blood Pressure: The JNC 7 Report. JAMA. 2003;289:2560-71.

60. Grundy SM, Cleeman JI, Merz CN, et al. Implications of recent clinical trials for the National Cholesterol Education Program Adult Treatment Panel III guidelines. *Circulation.* 2004;110:227-39.

Chapter 11
Cardiovascular Disease

Eunice P. Chung

Cardiovascular disease refers to all diseases of the circulatory system including hypertension, coronary heart disease, congestive heart failure, stroke, and congenital cardiovascular defects. It is the number one killer in the United States, claiming 1 of every 2.6 deaths.[1] One of the most common myths about heart disease is that it is a man's disease. Cardiovascular disease affects both men and women. In the last two decades, cardiovascular disease has claimed the lives of more females than males.

KEY POINTS

- Vitamin E supplementation does not appear to prevent cardiovascular disease or improve cardiovascular health.
- The American Heart Association recommends consuming a diet high in food sources of antioxidants.
- Epidemiological and observational evidence supports moderate alcohol consumption to reduce the risk of cardiovascular morbidity and mortality.
- Persons who do not normally drink alcohol should not be advised to start consumption for perceived cardiovascular benefit.
- Risks due to excessive alcohol consumption are well known.
- Patients with coronary heart disease should consume about 1 g of EPA and DHA combined per day, preferably from oily fish.
- Consumption of fish that contain large concentrations of contaminants should be limited in everyone and should be avoided in children and pregnant and lactating women.
- Persons without documented coronary heart disease can be counseled to consume a variety of fish at least twice a week to help preserve healthy cardiac function.
- The effect of garlic on blood pressure is small and is complicated by variances in product quality.
- Coenzyme Q10 may improve quality of life in patients with heart failure.
- Coenzyme Q10, hawthorn, and L-arginine may be useful as adjunctive therapy in chronic heart failure; however, present data are inconclusive.
- Currently, well-designed trials are underway to study the use of coenzyme Q10 and hawthorn in heart failure.
- L-Carnitine improves exercise capacity in heart failure patients with preserved cardiac function.

Product	Dosage	Effect	Safety Concerns
Vitamin E	Up to 400 IU/day	No benefit on cardiovascular disease	• Minimal gastrointestinal side effects • Bleeding risk with high doses • Increased risk of all-cause mortality at doses > 400 IU/day
Alcohol/wine	One or two drinks/day (men); one drink/day (women)	Decreases coronary heart disease	• Excessive use associated with many health and societal problems; refer to text
Fish oil (Omega-3 fatty acids)	1 g/day (EPA + DHA)	Decreases cardiovascular disease; decreases sudden cardiac death; decreases cardiac death; decreases blood pressure	• Nausea and loose stools at higher doses • Halitosis, belching • Can increase bleeding risk at higher doses (> 3 g/day)—use caution with concurrent use of antiplatelet/anticoagulant drugs and natural products that increase bleeding risk; monitor carefully for signs and symptoms of bleeding
Garlic	600 to 900 mg/day (dried garlic preparation)	Modest reduction in blood pressure	• Can increase bleeding risk • Malodorous breath or body odor
Coenzyme Q10	100 to 200 mg/day	Decreases blood pressure; improvement in heart failure symptoms and markers	• Rare increases of liver enzymes • Avoid coadministration with doxorubicin • Simvastatin, lovastatin, and gemfibrozil may decrease coenzyme Q10 levels
Hawthorn	180 to 1800 mg/day	Improves exercise capacity; improves heart failure symptoms	• Caution with concurrent use of digoxin

Product	Dosage	Effect	Safety Concerns
L-Arginine	3 to 21 g/day	Improves endothelial function; improves exercise capacity; prevents nitrate tolerance	• May contribute to hypotension and may aggravate hypotensive effects of other drugs • Higher doses have more risk of hypotensive effects • Do not use after a heart attack
L-Carnitine	2 g/day	Improves exercise capacity in patients with heart failure and peripheral artery disease; decreases left ventricular remodeling after heart attack	• May cause fishy odor in urine, breath, and sweat • Gastrointestinal effects (uncommon)

Coronary heart disease, which refers to myocardial infarction (heart attack) and angina pectoris (chest pain), accounts for more than half of cardiovascular disease-related deaths and is also the leading cause for heart failure. Significant efforts are devoted to identifying and treating the risk factors for coronary heart disease. Some known risk factors are high cholesterol, high blood pressure, diabetes, previous heart attack or stroke, age (> 45 for men and > 55 for women), family history (heart attack at age < 55 for male family members and < 65 for female family members), tobacco use, physical inactivity, and obesity. These risk factors are used to assess the likelihood of developing the first or recurrent heart attack. The risk for heart attack is greater for individuals who already experienced a heart attack.

It is common to encounter individuals who are reluctant to take prescription medication because they are asymptomatic and perceive themselves as healthy. Many of these individuals, however, are open to taking herbal or other dietary supplements. These products offer the convenience of avoiding physicians' office visits and costly prescriptions. Many consumers interpret "natural" products as safer alternatives to conventional drugs. Medications or dietary supplements are never a substitute for eating a healthy well-balanced diet.

Vitamin E

Vitamin E is a fat-soluble vitamin with antioxidant properties. According to the "oxidative-modification hypothesis," oxidation of low-density lipoprotein (LDL) plays an important role in the development of atherosclerosis.[2] The results of the early observational studies showing the association between vitamin E supplementation and reduced risk of coronary heart disease were attributed to the antioxidant effects of vitamin E.[3,4] Since then, several large-scale prospective studies have been conducted to investigate the true effects of vitamin E in preventing coronary heart disease. A few studies showed that vitamin E supplementation reduces the incidence of non-life-threatening heart attack.[5,6] However, the majority of studies failed to demonstrate any benefit in preventing heart disease. None of the studies showed a reduction in cardiovascular and noncardiovascular death rates.[7-11]

Vitamin E exists in at least eight naturally occurring compounds, of which α-tocopherol is the most active and prevalent compound. Chemical composition of naturally occurring α-tocopherol is different from the synthetic α-tocopherol. Both natural and synthetic vitamin E supplementation failed to demonstrate any benefit in preventing heart disease. Based on current findings, the American Heart Association (AHA) does not recommend antioxidant supplements such as vitamin E to reduce the risk of cardiovascular disease.[12] The recommendation is consistent for individuals who do not have diagnosed coronary heart disease and for individuals with known coronary heart disease.

The discrepancy between the observational studies and the controlled clinical trials are difficult to explain at this point. It could be that the duration of the controlled trials was not long enough to see the protective effect of vitamin E (they averaged 3 to 5 years), or it could be that the individuals did not start vitamin E supplementation early enough. While vitamin E supplementation is not recommended, the AHA does recommend consuming a diet high in food sources of antioxidants.[12] Good dietary sources of vitamin E include avocados, dark green leafy vegetables, eggs, nuts, peanut butter, and whole-grain breakfast cereals.[13]

Safety Considerations/Drug Interactions

Vitamin E supplements are sold in doses strikingly higher (400 to 1200 IU/day) than the recommended daily dietary allowance (15 IU for men and 12 IU for women). Previously, these high doses were considered safe and well tolerated. A recent meta-analysis concluded that high doses (> 400 IU/day) put patients at an increased risk of death.[14] In an earlier study, an increased incidence of hemorrhagic stroke raised concern regarding this potentially serious side effect.[15] However, this toxicity has not been reproduced in subsequent studies, even in those using higher doses of vitamin E.

High-dose vitamin E has been shown to display both antiplatelet and anticoagulant effects.[16,17] Vitamin E supplementation is not recommended for patients on anticoagulant therapy (e.g., warfarin) and should be used with caution for patients on antiplatelet therapy (e.g., aspirin or clopidogrel). Nevertheless, vitamin E supplements have been used for years in patients taking antiplatelet therapy. The significance of this drug interaction is yet to be determined. It may present more of a risk when elderly patients are using more than one agent that can increase bleeding risk, such as ginkgo biloba, ginseng, garlic, fish oils, and others. Many of the supplements that can increase bleeding risk are precisely those used by elderly patients.

ALCOHOL/WINE

There is considerable epidemiological and observational evidence to suggest that moderate consumption of alcohol lowers the risk of cardiovascular morbidity and mortality. Does it make sense to advise a patient to consume alcohol on a more regular basis? The answer is controversial because the current data are sufficient to establish a correlation but insufficient to prove a cause-and-effect relationship. A long-term prospective trial will provide a definitive answer. However, it is unlikely that such a trial will be conducted for various reasons, including ethical issues.

PATIENT CASES

PATTY—A HEALTHY PATIENT WITH ELEVATED BLOOD PRESSURE

Patty is a 36-year-old single female accountant with no specific medical complaints. Other than feeling tired from working long hours, Patty considers herself to be in a good state of health. She does not smoke and drinks alcohol only for rare social occasions. Currently, she is not on any chronic medications. On her last annual physical examination, Patty was told that everything is normal except for her elevated blood pressure. It was 145/94 mmHg. Rather than prescribing a medication to lower the blood pressure, her doctor recommended that she exercise and change her diet. Patty is slim and cannot find the time to exercise, but she does try to eat healthier food than before in light of her family history. Her father had a heart attack requiring open heart surgery at age 49 and a second operation again at age 60. Patty's mother is alive and well at age 67 with no chronic medical conditions.

Patty states that trying to eat healthy is a struggle for her because she eats most of her meals outside. While she does not like the idea of taking

PATIENT CASES continued

prescription medication for her blood pressure, she is interested in finding out if there are any natural products that could help her stay healthy. Ideally she would like a product that will control her blood pressure so that she could take a breather on trying to eat healthy all the time and that will lower her risk of getting a heart attack like her father.

BUDDY—A PATIENT WITH AN EXTENSIVE HISTORY OF HEART DISEASE

Buddy is a 65-year-old man with a long history of heart disease. Buddy was diagnosed with high blood pressure and high cholesterol more than 25 years ago, but because he was not feeling sick, he stopped taking his blood pressure and cholesterol-lowering medications soon after the diagnosis.

Buddy experienced his first heart attack at age 49, which required him to have open heart surgery. At age 57, he had another major heart attack, which forced him to retire early from his job as a restaurant manager. Buddy has since been diagnosed with heart failure. His most recent ejection fraction was 20%. He quit smoking and stopped drinking after his first heart attack. He lives with his wife who makes sure that he takes all his medications.

Despite these efforts, Buddy is continuously getting worse. These days, his daily activities are limited to staying in bed or sitting on the couch watching TV. Because his shortness of breath gets worse when lying down, he has to sleep with his head elevated using two or three pillows. Buddy's wife wants to know why he is getting worse when he takes all his medications religiously and also avoids eating unhealthy food. His current medications include aspirin, furosemide, digoxin, ramipril, carvedilol, spironolactone, amlodipine, and atorvastatin. His wife has seen advertisements of various dietary supplements that claim to rejuvenate the heart and is contemplating ordering them.

Several different mechanisms are thought to be involved in the protective effects of alcohol. Alcohol consumption has been shown to increase HDL cholesterol (the good cholesterol),[18,19] and up to half of the beneficial effects of alcohol have been attributed to the increase in HDL cholesterol.[20] Other major contributing effects are thought to be caused by alcohol's fibrinolytic (clot dissolving) effects and anti-platelet effects, similar to the cardio-protective effects of aspirin.[18,19,21]

Wine, in moderation, can be heart-healthy

There is an ongoing debate on whether the type of alcohol consumed affects the outcome, particularly whether red wine provides more benefit than other alcohols. The lower rate of cardiovascular disease in France compared to the United States, despite similar patterns of high fat food consumption, is referred to as the "French Paradox." The difference was attributed to red wine consumption by the French, with a subsequent hypothesis that red wine is superior to other types of alcohol in cardioprotection. It also raised the question whether drinking alcohol with a meal provides additional benefit since much of red wine is consumed with meals. A recent study comparing the effects of different alcohols (beer, red wine, white wine, and liquor) demonstrated that neither the type nor the consumption with meals altered the association.[22]

The relationship between alcohol consumption and mortality is described as a J-curve, where a moderate intake of alcohol is associated with a lower mortality rate and a heavier alcohol intake is associated with significant increases in mortality rate.[23,24] A moderate intake is generally defined as one or two drinks per day (one drink is equivalent to one 12-ounce beer, 4 ounces of wine, 1.5 ounces of 80-proof spirits, or 1 ounce of 100-proof spirits). Evidence from a meta-analysis indicates that moderate alcohol consumption reduces overall vascular risk by 32%, a number much too significant to ignore considering the burden of cardiovascular disease in public health.[23] However, because of the proven harmful effects of heavy alcohol consumption, it is difficult to publicly advocate or encourage alcohol drinking.

The American Heart Association recommends that patients who drink alcohol should be advised to do so in moderation (one or two drinks per day for men and one drink per day for women), but it does not recommend non-alcohol-drinking patients to start drinking to gain the potential cardiovascular benefits.[25] It also states that there is no clear evidence that wine is more beneficial than other forms of alcohol. For an occasional social drinker, it is not necessary to change the level of alcohol consumption.

Safety Considerations/Drug Interactions

Long-term and chronic heavy alcohol use (> three drinks per day) causes many adverse health effects, including liver cirrhosis, pancreatitis, hypertension, hemorrhagic stroke, cardiac arrhythmia, fetal alcohol syndrome, and sudden death.[26] In addition, alcohol is a proven addictive substance and a significant cause for automobile accidents, trauma, and suicide. Alcohol should not be considered as a preventive measure for teenagers and patients with uncontrolled hypertension, heart failure, pancreatitis, liver disease, and hypertriglyceridemia.[20,26]

Although heavy alcohol consumption is subject to many clinically significant drug interactions, light-to-moderate consumption is subject to minimal drug interactions. The U.S. Food and Drug Administration warns regular aspirin takers against drinking alcohol due to the potential additive antiplatelet effect, which may result in a higher incidence of bleeding.[25] A disulfiram reaction, a very unpleasant reaction caused by the accumulation of an alcohol metabolite, can occur with even small amounts of alcohol. Therefore, disulfiram (Antabuse®) and drugs that are thought to sometimes cause disulfiram-like reactions (e.g., griseofulvin and metronidazole) should not be used with alcohol.

Fish Oil (Omega-3 Fatty Acids)

Omega-3 fatty acids are essential fatty acids that must be obtained through the diet. Omega-3 fatty acids can be further differentiated into marine-derived, eicosapentaenoic acid (EPA) and docosahexaenoic acid (DHA), and plant-derived, alpha-linolenic acid (ALA), the precursor to EPA and DHA. Since the observation of a low incidence of coronary heart disease in some Eskimo populations despite a high-fat dietary pattern, a potential cardioprotective effect of omega-3 fatty acid consumption has been a subject of intense research. While both plant- and marine-derived omega-3 fatty acids have been shown to be beneficial to the heart, data for fish oil (i.e., EPA and DHA) are much more substantial.

Data to date support fish oil consumption for reducing heart attacks, sudden cardiac death, and the overall cardiac death rate. The magnitude of reduction is similar to the rates achieved with prescription HMG CoA reductase inhibitor ("statin") therapy.[27] These benefits are achievable with a modest dose of 1 g/day of EPA and DHA combined.[28]

There are several proposed mechanisms on how fish oil reduces cardiovascular morbidity and mortality. The positive effect of omega-3 fatty acids on triglyceride levels has been well established and is covered extensively in Chapter 10 of this book. Omega-3 fatty acids have been shown to have some antiarrhythmic properties by stabilizing the electrical activity of the heart. This property is hypothesized to be the main mechanism in decreasing the incidence of sudden cardiac death. A small, dose-dependent, blood pressure lowering effect has also been documented.

However, because the antihypertensive effects require high doses (> 3 g/day) for a small benefit and that there are many effective antihypertensive medications with mortality data to support their use, it is not a reasonable substitution to use fish oils for management of hypertension.[29] Omega-3 fatty acids have also been shown to inhibit atherosclerotic plaque formation. This antiatherogenic effect is attributed to the anti-inflammatory properties of omega-3 fatty acid and the ability to inhibit platelet aggregation. Omega-3 fatty acids improve endothelial function by enhancing the vasodilatory effect of nitrous oxide.

Current data provide sufficient evidence to support a significant role for fish oil in the secondary prevention of coronary heart disease. The American Heart Association recommends that patients with documented coronary heart disease consume about 1 g of EPA and DHA combined per day, preferably from oily fish.[29] All fish contain EPA and DHA, but the amount varies, depending on the species and the environmental factors. Fried fish from restaurants and frozen fish are not recommended because they are low in omega-3 fatty acids and can be high in saturated fats.

If regular fish consumption is not feasible, fish oil supplements can be considered. When educating patients on fish oil supplements, it is important to emphasize that the recommended daily dose of 1 g refers to EPA and DHA amounts rather than the dose of the fish oil. For example, a typical 1-g fish oil capsule could contain 180 mg of EPA and 120 mg of DHA. About three of these capsules would need to be taken on a daily basis to obtain the potentially beneficial cardiovascular effects.

For patients without documented coronary heart disease, a sound recommendation is to eat a variety of fish at least twice a week and to use ALA enriched oil or margarine (e.g., flaxseed, canola, and soybean oil) as substitutes for existing cooking oils and salad dressings. Supplementation with fish oil capsules, however, is not recommended for patients without preexisting coronary heart disease, until further evidence is available.[29]

Safety Considerations/Drug Interactions

Omega-3 fatty acids are a natural part of the human diet and are free of major side effects. The U.S. Food and Drug Administration has ruled that an intake of up to 3 g/day of fish oil is generally recognized as safe. An unpleasant fishy aftertaste may be experienced at higher doses but may not occur at cardioprotective doses. There is a potential for increased risk of bleeding when combined with drugs like warfarin or aspirin, but the evidence for bleeding has not been documented with daily doses less than 3 g.

Concerns for exposure to environmental contaminants (e.g., methylmercury, polychlorinated biphenyls, and dioxins) associated with increased fish consumption have been addressed recently. All persons should attempt to limit consumption of fish that have high levels of methylmercury. These include swordfish, shark, king mackerel, and tilefish. Children and pregnant and lactating women should take extra precautions to avoid these fish. For middle-aged and older men and postmenopausal women, the benefits of fish consumption outweigh the risks of exposure to environmental contaminants.[29]

Garlic

Garlic contains numerous sulfur-containing compounds, amino acids, vitamins, and minerals.[30] Allinase enzyme converts alliin to allicin when garlic is chopped or crushed. Allicin is thought to be the bioactive compound.[31] Garlic is primarily known for its cholesterol-lowering effect, which is covered in Chapter 10. Antithrombotic and blood pressure lowering are other cardiovascular effects attributed to garlic.

Antithrombotic effects consist of fibrinolytic activity (dissolving clots) and antiplatelet properties. The clinical role of garlic as an antithrombotic agent has not been established due to varying results from clinical trials. The antiplatelet effect does appear to be dose-dependent and it is thought that different components in garlic inhibit different stages of platelet aggregation.[31,32]

The blood pressure lowering effects of garlic have been shown to be modest at best. Only a fraction of the studies showed some benefit in lowering blood pressure.[33] In addition, most studies were not designed to compare blood pressure lowering effect as the primary outcome, and the studies were too small in sample size to draw meaningful conclusions. A report from the Agency for Healthcare Research and Quality, based on a systematic review and analysis of scientific evidence of garlic, indicated that there is no evidence that garlic has a beneficial impact on blood pressure.[34]

Current research also does not sufficiently address the variations in different garlic preparations. Garlic can be administered as fresh raw garlic, garlic powders, dried garlic tablets, aged garlic extract, or garlic oil. Garlic powder or the dried garlic tablets are considered to be virtually identical to fresh garlic in their chemical constituents, whereas the aged garlic extracts are thought to have different bioactive compounds. There are some limited data to suggest that aged garlic extract may lower blood pressure; however, the majority of the data on blood pressure lowering comes from dried garlic preparations of 600 to 900 mg/day.

Garlic bulbs

Patients may experience a slight lowering of blood pressure if garlic supplements are taken chronically. However, the weight of the evidence does not support garlic as an effective antihypertensive agent for routine clinical use at this point, particularly because there are many prescription antihypertensive agents with proven long-term safety and efficacy.

SAFETY CONSIDERATIONS/DRUG INTERACTIONS

The integral role garlic plays in the daily diet is a testament to its relative safety. Malodorous breath and body odor are the only side effects of garlic with clear causality.[35] Gastrointestinal discomfort, nausea, dermatitis, and rhinitis are other potential side effects of garlic. Bleeding is the most serious, potential side effect. An increased risk of bleeding was not observed in most studies. However, there was a reported case of epidural hematoma associated with a high dose of garlic intake.[36] Concomitant administration with other drugs that increase risk of bleeding (e.g., aspirin or warfarin) should be monitored carefully.

COENZYME Q10 (COQ10)

CoQ10 is a fat-soluble vitamin-like substance found in all tissues of the body, thus referred to as ubiquinone, with the highest concentrations in the heart, liver, kidneys, and pancreas. CoQ10 is necessary for certain metabolic reactions such as oxidative respiration, making it essential for production of adenosine triphosphate (ATP).[37,38] Biochemical findings indicate that CoQ10 levels decline in blood and myocardial tissue of failing hearts, and the reduction in concentration correlates with increasing severity of the disease. Subsequently, it was found that oral supplementation of CoQ10 is effective in increasing serum and tissue levels. CoQ10 can be synthesized within the body and is also found in food, primarily in meat, poultry, and fish.

CoQ10 supplementation has been studied for several cardiovascular conditions, including hypertension, cardiomyopathy, and ischemic heart disease, and, most extensively, for the treatment of chronic heart failure. There are two proposed mechanisms for its effect on the heart which includes an antioxidant effect and the ability to assist with more efficient usage of energy by the heart muscles during stress. These effects are thought to be optimally achieved at CoQ10 blood levels two to four times above the normal levels. These levels require supplementation beyond the daily dietary intake.[39]

Many studies have been published providing conflicting results regarding the efficacy of CoQ10 in the treatment of heart failure. Several studies that showed benefit used improvement in NYHA classification and left ventricular function as outcomes and demonstrated effectiveness of CoQ10 for treatment of heart failure at daily doses of 100 to 200 mg.[40-43] These results were refuted by other heart failure studies showing no improvement using similar outcome measurements.[44-47] However, there was less disparity in the overall quality of life outcome. Improved quality of life, measured by the symptomatic and exercise tolerance improvements, was demonstrated more consistently throughout the studies.

A meta-analysis of eight studies on treatment of chronic heart failure concluded that coenzyme Q10 supplementation is beneficial for treatment of chronic heart failure.[48] Another comprehensive review of coenzyme Q10 clinical studies on heart failure indicated that there is a positive trend for treatment of advanced heart failure.[49] Unfortunately, many of the studies have methodological flaws (e.g., lack of controls and blinded randomization and small sample sizes resulting in insufficient power) making it difficult to validate the results.

Therefore, it is premature to state conclusively whether CoQ10 is effective or not effective for treatment of heart failure. A large prospective study that will provide more definitive answers is currently underway. In the meantime, CoQ10 may be used as an adjunctive therapy for chronic heart failure under the supervision of a physician but should not be used in place of conventional heart failure medications.

Safety Considerations/Drug Interactions

At therapeutic doses of 100 to 200 mg/day, CoQ10 is devoid of any major side effects. A few minor side effects such as epigastric discomfort, nausea, diarrhea, and rare increases of liver enzymes have been reported that could possibly be related to CoQ10 therapy.[49]

Concomitant administration of CoQ10 and the chemotherapy drug doxorubicin (Adriamycin®) should be avoided as CoQ10 can alter the metabolism of doxorubicin and increase the concentration of a potentially toxic metabolite. CoQ10 therapy, however, has been shown to be effective in preventing cardiac toxicities of doxorubicin, when used after the cessation of chemotherapy.

Cholesterol-lowering drugs such as statins (e.g., simvastatin [Zocor®], lovastatin [Mevacor®]) and gemfibrozil (Lopid®) may decrease plasma and tissue CoQ10 levels. It is unclear whether normalizing coenzyme Q10 levels via supplementation will benefit patients on statins and gemfibrozil. Pravastatin (Pravachol®) and atorvastatin (Lipitor®) do not lower coenzyme Q10 levels.[50] The beta-blockers propranolol and metoprolol may also inhibit coenzyme Q10-dependent enzymes and ultimately lower CoQ10 levels. CoQ10 is structurally similar to vitamin K. Therefore, a procoagulant effect (resulting in a decreased international normalized ratio) when combined with warfarin has been suggested, but a small study found no interaction between CoQ10 and warfarin.[51]

HAWTHORN

Hawthorn is a common name for all fruit-bearing plants in the Crataegus species. Extracts from its leaves, flowers, and berries have been used for various medicinal purposes for centuries in Europe and Asia. Although there are many chemical constituents, flavonoid and oligomeric proanthocyanidins (OPCs) are considered to be the main active components for effects on the cardiovascular system.[52,53] Therefore, many hawthorn preparations are standardized based on their flavonoid and OPC content (2.2% of flavonoids or 18.75% of OPC).[53] As a cardiovascular agent, hawthorn is used for the treatment of angina, hypertension, heart failure, and arrhythmias. However, most available human studies relate to its use in the treatment of mild heart failure or those in NYHA class II heart failure. The majority of these studies were conducted in Germany, where hawthorn preparations are marketed as prescription medications for the treatment of mild heart failure.

The pharmacologic properties of hawthorn on the cardiovascular system are attributed to a positive inotropic effect, coronary and peripheral vasodilation, a protective action against ischemia-induced ventricular arrhythmias, and antioxidant properties.[52-54] The available data are promising for NYHA class II heart failure patients in improving exercise capacity, subjective symptoms of heart failure, and the blood pressure/heart rate product, which correlates with the oxygen consumption of the heart.[55]

Previously it was thought that only the hawthorn leaf and flower extracts were useful for treating heart failure, but a recent study demonstrated that hawthorn berry extracts provide similar benefits.[56] The use of hawthorn may also expand to include patients with more severe heart failure. A randomized controlled study showed that a high dose of hawthorn extract (1800 mg/day) is beneficial for treating NYHA class III patients.[57]

It is important to keep in mind that the outcomes evaluated in these studies are surrogate endpoints measured in relatively small group of patients over a short period of time. A long-term study using mortality endpoints is critical for understanding the long-term safety and survival benefit. A study currently underway is expected to provide these answers and further establish hawthorn's role in the treatment of heart failure.[58]

Hawthorn extracts are available in capsules, liquids, and teas. Doses for leaf and flower extracts tested in clinical studies range from 160 to 1800 mg/day, while the berry extracts were tested from 300 to 1000 mg three times a day. Hawthorn supplements should not be used as a replacement for standard heart failure therapy.

SAFETY CONSIDERATIONS/DRUG INTERACTIONS

Hawthorn was well tolerated in studies lasting up to 16 weeks. Some side effects that were rare but may be related to hawthorn extracts were mild rash, headache, sweating, dizziness, sleepiness, agitation, and gastrointestinal complaints.[52,53,55] Drug interaction profiles are not well studied, so most drug interactions are theoretical based on the pharmacologic properties. Most hawthorn products warn against using this product concomitantly with digoxin due to the perceived potentiation of digoxin's effects. This interaction was refuted in one study which showed that hawthorn does not significantly alter digoxin pharmacokinetic parameters.[59] However, digoxin should still be monitored closely if the decision is made to take hawthorn concomitantly. Although not proven, hawthorn may interact with other vasodilating medications that are used for treatment of hypertension, angina, and heart failure.

L-ARGININE

The amino acid L-arginine serves as a substrate for the enzyme nitric oxide synthase, which converts L-arginine to nitric oxide. Nitric oxide, also referred to as the endothelium-derived relaxing factor (EDRF), causes vasodilation of vascular smooth muscle. It also plays a role in platelet aggregation and adhesion. In patients with coronary artery disease or its risk factors, the endothelial release of nitric oxide is reduced or absent. The reduced or lack of nitric oxide compromises the blood flow during myocardial stress and may contribute to ischemia. L-Arginine supplementation is hypothesized to improve the endothelial function in patients with cardiovascular disease by restoring the nitric oxide effects.

Early human studies were conducted with a single dose of intravenously administered L-arginine. While the intravenous L-arginine studies provided uniformly positive results, administering intravenous L-arginine supplementation is not practical. The oral L-arginine studies have provided inconsistent results. Many studies showed that oral L-arginine improves peripheral blood flow, as measured by flow-mediated vasodilation, and improved exercise capacity.[60-62] L-Arginine was also shown to prevent or modify the development of nitrate tolerance when used in conjunction with transdermal nitroglycerin in angina patients.[63] Contrary to some findings,[60-63] oral L-arginine supplementation showed no beneficial effect when combined with optimal medical management in patients with chronic stable angina.[64] The exact reason for the inconsistent results is unknown, but one possibility is the baseline differences in the studied population. Current data suggest that patients with hypercholesterolemia tend to benefit more from L-arginine supplementation than patients with other cardiovascular disease risk factors. The clinical applicability at this point is limited by the inconsistent results and the intermediate endpoints (e.g. flow-mediated vasodilation) used in the studies. Larger clinical trials with clinically applicable endpoints (e.g., reduction of coronary heart disease or death rates) are needed to determine the role of L-arginine supplementation in cardiovascular disease prevention or treatment.

L-Arginine is available as dietary supplements in capsules and tablets and was available as a medical food called HeartBar (now discontinued). HeartBar contained more L-arginine (3.3 g) than most L-arginine dietary supplements (100 to 500 mg). HeartBar also contained many other vitamins and minerals and was promoted for the dietary management of vascular disease.[65] Clinical trials using HeartBar showed conflicting results. One study showed no favorable effect on endothelial function or platelet function,[66] while another study demonstrated significant improvement in exercise capacity and quality of life.[67] Meat, fish, eggs, peanuts, and dairy products are rich in L-arginine.

SAFETY CONSIDERATIONS/DRUG INTERACTIONS

L-Arginine was well tolerated in clinical trials. The most common side effects were infrequent nausea, diarrhea, and stomach cramps, especially at higher doses. Patient compliance with higher doses may be poor due to slightly bitter and unpleasant taste of L-arginine.[65] At a recommended consumption of two HeartBars per day, fat content (3 g/bar) may be problematic for patients with cardiovascular disease who should limit overall fat intake.

Due to L-arginine's vasodilatory effects, it has the potential to lower blood pressure and potentially worsen hypotension. Patients with hypotension episodes should avoid L-arginine. Due to L-arginine's vasodilatory effects, it might have additive blood pressure lowering effects when combined with antihypertensives, nitrates, and phosphodiesterase inhibitors such as sildenafil (Viagra®).

CARNITINE

L-Carnitine can be produced endogenously in the liver and kidneys using the amino acids lysine and methionine. Therefore, it is not essential for most individuals to consume L-carnitine through the diet. Red meats are the best sources of L-carnitine, but it can also be found in fish, chicken, and dairy products. L-Carnitine transports long-chain fatty acids to the mitochondria where they are burned for energy production. Experimental studies have shown that there is depletion of carnitine in heart tissues during heart attacks and in patients with heart failure. These levels can be restored with exogenous L-carnitine administration.[68-70]

L-Carnitine has been shown to increase exercise tolerance and capacity in patients with heart failure or peripheral artery disease.[70-73] However, a relatively large double-blind, randomized clinical trial demonstrated that L-carnitine improves exercise capacity only in heart failure patients with preserved cardiac function (ejection fraction > 30%) and provided no beneficial effect in patients with severe heart failure.[74] Administration of L-carnitine following an acute heart attack causes attenuation of left ventricular dilation and results in smaller ventricular volumes.[68,75] Larger ventricles are associated with higher risks of complications from heart attacks, such as heart failure, arrhythmia, and death. Although it remains to be proven, reduction in ventricular volume is expected to ultimately reduce these complications.

SAFETY CONSIDERATIONS/DRUG INTERACTIONS

Side effects or drug interactions were rarely discussed in cardiovascular studies of L-carnitine. A few side effects that were reported include nausea, vomiting, stomach upset, diarrhea, and weakness, but none of the side effects was serious enough to cause discontinuation of L-carnitine administration. L-Carnitine may cause a fishy odor of the urine, breath, and sweat.[76]

PATIENT DISCUSSION

Patty is still young (< 55 for women), does not have diabetes or high cholesterol, does not use tobacco, has never experienced coronary heart disease, and is slim. However, her blood pressure is mildly elevated, she lives a sedentary lifestyle, and, most importantly, she has a strong family history (father had a heart attack at age < 55). Although her absolute risk of developing coronary heart disease in the next 10 years is very low (< 5%), it is important for Patty to work on her modifiable risk factors such as her blood pressure and physical inactivity, given the significant nonmodifiable risk factor of family history. Patty's blood pressure would be classified as stage 1 hypertension. With lifestyle modification and/or drug therapy, she should be able to control her blood pressure to less than 140/90 mmHg. She should ultimately aim to reach and maintain a normal blood pressure of less than 120/80 mmHg. A minimum of 30 minutes of physical activity on most days of the week will be important to control her blood pressure as well as to condition her heart and lungs.

Buddy, however, has several major risk factors for coronary heart disease, including hypertension, hyperlipidemia, and smoking. Like Buddy, many patients ignore these controllable risk factors due to the absence of symptoms until it is too late. He has survived two major heart attacks and is now suffering from chronic heart failure, the most common sequalae of a heart attack. Unfortunately there is no cure for heart failure. The heart muscles have stretched and become too weak to pump and supply blood to the rest of the body. Buddy is receiving optimal drug therapy, but unfortunately, the current drug therapy cannot prevent or reverse the progression of the disease. It will only slow the rate of disease progression. The mortality rate of heart failure is very high, up to 80% within 5 years of diagnosis.

Considering his significant cardiac history and knowing that he has been diagnosed with heart failure for 8 years now, Buddy's disease progression is consistent with epidemiologic data. His low ejection fraction and limitations on daily activities suggest that his heart failure is severe and can be classified as New York Heart Association (NYHA) class IV. Unfortunately, there are no additional prescription medications that can be prescribed to him for chronic management of his heart failure. It is interesting to consider whether he might benefit from any of the dietary supplements that are being promoted for cardiovascular disease.

Although preliminary data appear convincing for some of the natural products discussed here, the evidence is much less robust compared to data supporting the use of the prescription medications that Buddy is currently taking. Carvedilol, ramipril, and spironolactone all belong to different classes of medications that have been shown to increase survival rate in heart failure patients. It is critical that Buddy remains compliant with his current medication regimen. It is unlikely that he can benefit from any of the natural products discussed given the severity of his conditions. The majority of the beneficial effects of the natural products were documented in patients with mild-to-moderate heart failure or patients with preserved cardiac function. Since he is on multiple medications, the potential for interactions with natural products also must be considered (e.g., hawthorn and digoxin).

Self-Assessment

1. Which of the following statements correctly describe the findings of vitamin E supplementation on cardiovascular disease outcome:
   a. It lowers the incidence of heart attack and stroke.
   b. It provides protective effects against heart attack but not against stroke.
   c. It prolongs the survival of patients with coronary heart disease.
   d. It does not lower the incidence of coronary heart disease.

2. Which of the following correctly states the American Heart Association recommendation on alcohol intake for patients with coronary heart disease:
   a. All patients with coronary heart disease risk factors should abstain from drinking alcohol.
   b. Non-alcohol-drinking patients with coronary heart disease risk factors should start drinking alcohol.
   c. Patients who already drink alcohol should be advised to limit drinking to one or two drinks per day.
   d. Individuals who drink beer should be advised to drink red wine instead.

3. Scientific evidence suggests that about a gram of fish oil daily may be beneficial for patients with established coronary heart disease. If a product is labeled as 1 g of fish oil with 180 mg EPA and 120 mg of DHA per capsule, about how many capsules should be taken each day?
   a. One capsule
   b. Two capsules
   c. Three capsules
   d. Four capsules

4. Which of the following are considered to be the active ingredients in hawthorn?
   a. Flavonoid and oligomeric proanthocyanidins
   b. Alpha-linolenic acid
   c. Eicosapentaenoic acid and docosahexaenoic acid
   d. Crataegus

5. Which of the following is the most serious potential side effect of garlic?
   a. Rhinitis
   b. Bleeding
   c. Blood clot
   d. Infection

6. Which of the following is the proper recommendation for fish oil consumption:
   a. Patients with established coronary heart disease should consume one or two fish meals per week.
   b. Fried and frozen fish contain the same amount of omega-3 fatty acid as fresh fish.
   c. Fish oil supplements are preferable over the consumption of fresh fish due to ease of administration.
   d. Fish oil supplements should be recommended for primary prevention of coronary heart disease.

7. Which of the following natural products has been tested most extensively for treatment of heart failure?
   a. Garlic
   b. Vitamin E
   c. L-Arginine
   d. Coenzyme Q10

8. Which of the following groups of patients is most likely to benefit from hawthorn supplements:
   a. Healthy patients with inherent risk factors for developing heart failure.
   b. Patients experiencing acute heart attack.
   c. NYHA Class II heart failure patients.
   d. NYHA Class IV heart failure patients.

9. What is the active ingredient in the medical food, HeartBar?
   a. Coenzyme Q10
   b. L-Carnitine
   c. Hawthorn
   d. L-Arginine

10. Which of the following describes the proposed effect of L-carnitine when administered following an acute heart attack:
    a. It speeds up the recovery period.
    b. It minimizes left ventricular stretching.
    c. It reduces the need for supplemental oxygen.
    d. It lowers the blood pressure and the heart rate.

Answers: 1-d; 2-c; 3-c; 4-a; 5-b; 6-a; 7-d; 8-c; 9-d; 10-b

REFERENCES

1. American Heart Association. Heart disease and stroke statistics–2004 update. Dallas: American Heart Association; 2003.

2. Diaz MN, Frei B, Vita JA, et al. Antioxidants and atherosclerotic heart disease. *N Engl J Med.* 1997;337:408-16.

3. Rimm EB, Stampfer MJ, Ascherio A, et al. Vitamin E consumption and the risk of coronary heart disease in men. *N Engl J Med.* 1993;328:1450-6.

4. Stampfer MJ, Hennekens CH, Manson JE, et al. Vitamin E consumption and the risk of coronary disease in women. *N Engl J Med.* 1993;328:1444-9.

5. Stephens NG, Parsons A, Schofield PM, et al. Randomised controlled trial of vitamin E in patients with coronary disease: Cambridge Heart Antioxidant Study (CHAOS). *Lancet.* 1996;347:781-6.

6. Rapola JM, Virtamo J, Ripatti S, et al. Randomised trial of α-tocopherol and β-carotene supplements on incidence of major coronary events in men with previous myocardial infarction. *Lancet.* 1997;349:1715-20.

7. GISSI-Prevenzione Investigators. Dietary supplementation with n-3 polyunsaturated fatty acids and vitamin E after myocardial infarction: results of the GISSI-Prevenzione trial. *Lancet.* 1999;354:447-55.

8. Collaborative Group of the Primary Prevention Project. Low-dose aspirin and vitamin E in people at cardiovascular risk: a randomized trial in general practice. *Lancet.* 2001;357:89-95.

9. Hodis HN, Mack WJ, LaBree L, et al. Alpha-tocopherol supplementation in healthy individuals reduces low-density lipoprotein oxidation but not atherosclerosis: the vitamin E atherosclerosis prevention study (VEAPS). *Circulation.* 2002;106:1453-9.

10. The Heart Outcomes Prevention Evaluation Study Investigators. Vitamin E supplementation and cardiovascular events in high-risk patients. *N Engl J Med.* 2000;342:154-60.

11. Heart Protection Study Collaborative Group. MRC/BHF heart protection study of antioxidant vitamin supplementation in 20,536 high-risk individuals: a randomised placebo-controlled trial. *Lancet.* 2002;360:23-33.

12. Kris-Etherton PA, Lichtenstein AH, Howard BV, et al. AHA science advisory: antioxidant vitamin supplements and cardiovascular disease. *Circulation.* 2004;110:637-41.

13. Emmert DH, Kirchner JT. The role of vitamin E in the prevention of heart disease. *Arch Fam Med.* 1999;8:537-42.

14. Miller ER 3rd, Pastor-Barriuso R, Dalal D, et al. Meta-analysis: high-dosage vitamin E supplementation may increase all-cause mortality. *Ann Intern Med.* 2005;142:37-46.

15. The Alpha-Tocopherol, Beta Carotene Cancer Prevention Study Group. The effect of vitamin E and beta carotene on the incidence of lung cancer and other cancers in male smokers. *N Engl J Med.* 1994;330:1029-35.

16. Murohara T, Ikeda H, Otsuka Y, et al. Inhibition of platelet adherence to mononuclear cell by α-tocopherol: role of P-selectin. *Circulation.* 2004;110:141-8.

17. Booth SL, Golly I, Sacheck JM, et al. Effect of vitamin E supplementation on vitamin K status in adults with normal coagulation status. *Am J Clin Nutr.* 2004;80:143-8.

18. Gaziano JM, Buring JE, Breslow JL, et al. Moderate alcohol intake, increased levels of high-density lipoprotein and its subfractions, and decreased risk of myocardial infarction. *N Engl J Med.* 1993;329:1829-34.

19. Rimm EB, Williams P, Fosher K, et al. Moderate alcohol intake and lower risk of coronary heart disease: meta-analysis of effects on lipids and haemostatic factors. *BMJ.* 1999;319:1523-8.

20. Pearson TA. Alcohol and heart disease. *Circulation.* 1996;94:3023-5.

21. Ridker PM, Vaughan DE, Stampfer MJ, et al. Association of moderate alcohol consumption and plasma concentration of endogenous tissue-type plasminogen activator. *JAMA.* 1994;272:929-33.

22. Mukamal KJ, Conigrave KM, Mittleman MA, et al. Roles of drinking pattern and type of alcohol consumed in coronary heart disease in men. *N Engl J Med.* 2003;348:109-18.

23. Castelnuovo AD, Rotondo S, Iacoviello L, et al. Meta-anaylsis of wine and beer consumption in relation to vascular risk. *Circulation.* 2002;105:2836-44.

24. Bradley KA, Donovan DM, Larson EB. How much is too much? Advising patients about safe levels of alcohol consumption. *Arch Intern Med.* 1993;153:2734-40.

25. Alcohol, wine and cardiovascular disease. American Heart Association. Available at www.americanheart.org/presenter.jhtml?identifier=4422. Accessed October 6, 2004.

26. Goldberg IJ, Mosca L, Piano MR, et al. Wine and your heart: a science advisory for healthcare professionals from the Nutrition Committee, Council on Epidemiology and Prevention, and Council on Cardiovascular Nursing of the American Heart Association. *Circulation.* 2001;103:472-5.

27. Harper CR, Jacobson TA. The fats of life. The role of omega-3-fatty acids in the prevention of coronary heart disease. *Arch Intern Med.* 2001;161:2185-92.

28. Nestel PJ. Fish oil and cardiovascular disease: lipids and arterial function. *Am J Clin Nutr.* 2000;71 (Suppl):228S-31S.

29. Kris-Etherton PM, Harris WS, Appel LJ, et al. Fish consumption, fish oil, omega-3 fatty acids, and cardiovascular disease. *Circulation.* 2002;106:2747-57.

30. Brace LD. Cardiovascular benefits of garlic. *J Cardiovasc Nurs.* 2002;16(4):33-49.

31. Banerjee SK, Maulik SK. Effect of garlic on cardiovascular disorders: a review. *Nutrition J*. 2002;1:4.

32. Steiner M, Li W. Aged garlic extract, a modulator of cardiovascular risk factors: a dose-finding study on the effects of AGE on platelet function. *J Nutr*. 2001;13:980S-4S.

33. Silagy CA, Neil HA. A meta-analysis of the effect of garlic on blood pressure. *J Hypertension*. 1994;12:463-8.

34. AHRQ Report Finds Little or Inconclusive Evidence of Health Benefits of Garlic. Press Release, October 3, 2000. Agency for Healthcare Research and Quality, Rockville, MD. Available at www.ahrq.gov/news/press/pr2000/garlicpr.htm. Accessed October 21, 2004

35. Ackermann RT, Mulrow CD, Ramirez G, et al. Garlic shows promise for improving some cardiovascular risk factors. *Arch Intern Med*. 2001;161:813-24.

36. Rose KD, Croissant PD, Parliament CF, et al. Spontaneous spinal epidural hematoma with associated platelet dysfunction from excessive garlic ingestion: a case report. *Neurosurgery*. 1990;26:880-2.

37. Morelli V, Zoorob RJ. Alternative therapies: part II. Congestive heart failure and hypercholesterolemia. *Am Fam Physician*. 2000;62:1325-30.

38. Sarter B. Coenzyme Q 10 and cardiovascular disease: a review. *J Cardiovasc Nurs*. 2002;16(4):9-20.

39. Langsjoen PH, Langsjoen AM. Overview of the use of CoQ10 in cardiovascular disease. *BioFactors*. 1999;9:273-84.

40. Sacher HL, Sacher ML, Landau SW, et al. The clinical and hemodynamic effects of coenzyme Q10 in congestive cardiomyopathy. *Am J Ther*. 1997;4:66-72.

41. Langsjoen PH, Langsjoen AM, Willis R, et al. Treatment of hypertrophic cardiomyopathy with coenzyme Q10. *Mol Aspects Med*. 1997;18(Suppl):S145-51.

42. Munkholm H, Hansen HH, Rasmussen K. Coenzyme Q10 treatment in serious heart failure. *BioFactors*. 1999;9:285-9.

43. Morisco C, Trimarco B, Condorelli M. Effect of coenzyme Q10 therapy in patients with congestive heart failure: a long-term multicenter randomized study. *Clin Invest*. 1993;71(8 suppl):S134-6.

44. Hofman-Bang C, Rehnqvist N, Swedberg K, et al. Coenzyme Q10 as an adjunctive in the treatment of chronic congestive heart failure. The Q10 Study Group. *J Card Fail*. 1995;1:101-7.

45. Chen YF, Lin YT, Wu SC. Effectiveness of coenzyme Q10 on myocardial preservation during hypothermic cardioplegic arrest. *J Thorac Cardiovasc Surg*. 1994;107:242-7.

46. Watson PS, Scalia GM, Galbraith A, et al. Lack of effect of coenzyme Q on left ventricular function in patients with congestive heart failure. *J Am Coll Cardiol*. 1999;33:1549-52.

47. Khatta M, Alexander BS, Krichten CM, et al. The effect of coenzyme Q10 in patients with congestive heart failure. *Ann Intern Med.* 2000;132:636-40.

48. Soja AM, Mortensen SA. Treatment of congestive heart failure with coenzyme Q10 illuminated by meta-analyses of clinical trials. *Mol Aspects Med.* 1997;18(Suppl):S159-68.

49. Mortensen SA. Overview on coenzyme Q10 as adjunctive therapy in chronic heart failure. Rationale, design and end-points of "Q-symbio"—a multinational trial. *BioFactors.* 2003;18:79-89. Review.

50. Bleske BE, Willis RA, Anthony M, et al. The effect of pravastatin and atorvastatin on coenzyme Q10. *Am Heart J.* 2001;142:E2-7.

51. Engelsen J, Nielsen JD, Winther K. Effect of coenzyme Q10 and Ginkgo biloba on warfarin dosage in stable, long-term warfarin-treated outpatients. A randomised, double-blind, placebo-crossover trial. *Thromb Haemost.* 2002;87:1075-6.

52. Rigelsky JM, Sweet BV. Hawthorn: pharmacology and therapeutic uses. *Am J Health Syst Pharm.* 2002;59:417-22.

53. Chang Q, Zuo Z, Harrison F, et al. Hawthorn. *J Clin Pharmacol.* 2002;42:605-12.

54. Fong HHS, Bauman JL. Hawthorn. *J Cardiovasc Nurs.* 2002;16(4):1-8.

55. Pittler MH, Schmidt K, Ernst E. Hawthorn extract for treating chronic heart failure: meta-analysis of randomized trials. *Am J Med.* 2003;114:665-74.

56. Degenring FH, Suter A, Weber M, et al. A randomised double blind placebo controlled clinical trial of a standardized extract of fresh Crataegus berries (Crataegisan®) in the treatment of patients with congestive heart failure NYHA II. *Phytomedicine.* 2003;10:363-9.

57. Tauchert M. Efficacy and safety of crataegus extract WS 1442 in comparison with placebo in patients with chronic stable New York Heart Association class-III heart failure. *Am Heart J.* 2002;143:910-5.

58. Holubarsch CJ, Colucci WS, Meinertz T, et al. Survival and prognosis: investigation of crataegus extract WS 1442 in congestive heart failure (SPICE): rationale, study design and study protocol. *Eur J Heart Fail.* 2000;2:431-7.

59. Tankanow R, Tamer HR, Streetman DS, et al. Interaction study between digoxin and a preparation of hawthorn (Crataegus oxycantha). *J Clin Pharmacol.* 2003;43:637-42.

60. Hambrecht R, Hilbrich L, Erbs S, et al. Correction of endothelial dysfunction in chronic heart failure: additional effects of exercise training and oral L-arginine supplementation. *J Am Coll Cardiol.* 2000;35:706-13.

61. Rector TS, Bank AJ, Mullen KA, et al. Randomized, double-blind, placebo-controlled study of supplemental oral L-arginine in patients with heart failure. *Circulation.* 1996;93:2135-41.

62. Ceremuzynski L, Chamiec T, Herbaczynska-Cedro K. Effect of supplemental oral L-arginine on exercise capacity in patients with stable angina pectoris. *Am J Cardiol.* 1997;80:331-3.

63. Parker JO, Parker JD, Caldwell RW, et al. The effect of supplemental L-arginine on tolerance development during continuous transdermal nitroglycerin therapy. *J Am Coll Cardiol.* 2002;39:1199-203.

64. Blum A, Hathaway L, Mincemoyer R, et al. Oral L-arginine in patients with coronary artery disease on medical management. *Circulation.* 2000;101:2160-4.

65. Cheng JWM, Baldwin SN. L-Arginine in the management of cardiovascular diseases. *Ann Pharmacother.* 2001;35:755-64.

66. Abdelhamed AI, Reis SE, Sane DC, et al. No effect of an L-arginine-enriched medical food (HeartBars) on endothelial function and platelet aggregation in subjects with hypercholesterolemia. *Am Heart J.* 2003;145:E15.

67. Maxwell AJ, Zapien MP, Pearce GL, et al. Randomized trial of a medical food for the dietary management of chronic, stable angina. *J Am Coll Cardiol.* 2002;39:37-45.

68. Iliceto S, Scrutinio D, Bruzzi P, et al. Effects of L-carnitine administration on left ventricular remodeling after acute anterior myocardial infarction: the L-Carnitine Ecocardiografia Digitalizzata Infarcto Miocardico (CEDIM) trial. *J Am Coll Cardiol.* 1995;26:380-7.

69. Rizos, I. Three-year survival of patients with heart failure caused by dilated cardiomyopathy and L-carnitine administration. *Am Heart J.* 2000;139(2 Pt 3):S120-3.

70. Loster H, Miehe K, Punzel M, et al. Prolonged oral carnitine substitution increases bicycle ergometer performance in patients with severe, ischemically induced cardiac insufficiency. *Cardiovasc Drugs Ther.* 1999;13:537-46.

71. Anand I, Chandrashekhan Y, De Giuli F, et al. Acute and chronic effects of propionyl-L-carnitine on the hemodynamics, exercise capacity, and hormones in patients with congestive heart failure. *Cardiovasc Drugs Ther.* 1998;12:291-9.

72. Mancini M, Rengo F, Lingetti M, et al. Controlled study on the therapeutic efficacy of propionyl-L-carnitine in patients with congestive heart failure. *Arzneimittelforschung.* 1992;42:1101-4.

73. Hiatt WR, Regensteiner JG, Creager MA, et al. Propionyl-L-carnitine improves exercise performance and functional status in patients with claudication. *Am J Med.* 2001;110:616-22.

74. The Investigators of the Study on Propionyl-L-Carnitine in Chronic Heart Failure. Study on propionyl-L-carnitine in chronic heart failure. *Eur Heart J.* 1999;20:70-6.

75. Singh RB, Niaz MA, Agarwal P, et al. A randomised, double-blind, placebo-controlled trial of L-carnitine in suspected acute myocardial infarction. *Postgrad Med J.* 1996;72:45-50.

76. Evans AM, Fornasini G. Pharmacokinetics of L-carnitine. *Clin Pharmacokinet.* 2003;42:941-67.

Chapter 12
Diabetes

Winston Y. Wong

Type 2 diabetes is the most common form of diabetes, representing 90% to 95% of cases. The risk factors for type 2 diabetes include family history, metabolic syndrome, ethnicity (persons of African, Hispanic, Native American, Asian, or Pacific Island descent), age > 45 years, previously identified impaired glucose tolerance (IGT), a history of gestational diabetes mellitus, and an inactive lifestyle (exercise fewer than three times per week).

This chapter covers agents for improving blood glucose only. The chapters on cholesterol and cardiovascular disease discuss agents used for controlling cholesterol and blood pressure. These two areas are critically important in these patients to help reduce complications. As far as blood glucose is concerned, the American Diabetes Association recommends a fasting plasma glucose (FPG) goal between 90 and 130 mg/dL, a postprandial plasma glucose (PPG) level of less than 180 mg/dL, and a glycated hemoglobin (A1C) value of less than 7%.[1]

Patients who request natural products to lower blood glucose represent an opportunity to discuss the most important "natural" ways to manage this disease: physical activity and diet. This is largely a lifestyle disease for most patients. Currently many people are largely sedentary and subsist on highly processed, refined foods.

KEY POINTS

- Bitter melon, gymnema, prickly pear cactus, and American ginseng may cause hypoglycemia and should be used cautiously with hypoglycemic drugs.
- Chromium deficiency exists in some patients with diabetes.
- Certain patients, including those with known chromium deficiency, may benefit from chromium supplementation.
- Cassia cinnamon, in reasonable doses, helps lower blood glucose and improves lipids.
- Blond psyllium and prickly pear cactus can be used to help lower postprandial blood glucose levels.
- Prickly pear cactus, taken as a food form several times daily, is not practical for most patients and the dried extracts are not effective.
- American ginseng must be cultivated; wild American ginseng is endangered.

Initial physical activity recommendations should be modest, based on the patient's willingness and ability, with gradual increases in duration and intensity.[1] There are several recommended diet regimens for type 2 diabetes patients which focus on the need to increase the intake of complex carbohydrates rich in "dietary fiber" and water-soluble fiber such as legumes, oat bran, nuts, seeds, psyllium seed husks, pears, apples, and most vegetables. In addition, diabetes patients have an increased need for many nutrients such as vitamin C, niacin, biotin, vitamin B_6, vitamin B_{12}, vitamin E, magnesium, potassium, manganese, zinc, flavonoids, carnitine, chromium, inositol, lipoic acid, and essential fatty acids (i.e., omega-6 and omega-3 fatty acids).[2]

Product	Dosage	Effect	Safety Concerns
Bitter melon	No standard dose based on clinical studies	Several preliminary studies; efficacy unknown	• Possibility of hypoglycemia if combined with hypoglycemic drugs
Chromium	200 to 1000 mcg/day in divided doses	Decreases blood glucose and A1C in 40% to 80% of patients; decreases A1C ~0.6%; may work best in chromium-deficient patients	• Cognitive and mood changes; caution in patients with psychiatric disorders; significant adverse effects with very high doses; see text
Cassia cinnamon	1 to 6 g/day; 1 teaspoon is 4.75 g	Decreases fasting blood glucose up to 29%, decreases LDL and triglycerides	• None known
Gymnema	Specific gymnema extract (GS4) 400 mg/day	Two preliminary studies; efficacy unknown	• Possibility of hypoglycemia if combined with hypoglycemic drugs
Blond psyllium	15 g/day in divided doses	Decreases postprandial glucose by up to 20%; also benefit for lipids (see Cholesterol Reduction chapter)	• Avoid in patients with gastrointestinal conditions such as bowel obstruction or swallowing disorders; maintain adequate fluid intake • May bind to certain oral medication

Product	Dosage	Effect	Safety Concerns
Stevia	1 g/day of stevioside	One preliminary clinical study; efficacy unknown	• Cross-allerginicity with ragweed
Prickly pear cactus	100 to 500 g/day in divided doses (dose not practical); broiled stems only (not supplements)	Several preliminary studies; decreases blood glucose by 17% to 46%	• Possibility of hypoglycemia if combined with hypoglycemic drugs
American ginseng	3 g up to 2 hours before a meal	Several preliminary small clinical trials show decreases in postprandial glucose	• Possibility of hypoglycemia if combined with hypoglycemic drugs • May decrease warfarin effect • May potentiate effects of stimulant drugs, antipsychotic drugs, and monoamine oxidase inhibitors • Do not use in pregnancy

Bitter Melon

Bitter melon (Momordica charantia) is a popular vegetable in Southeast Asia. It looks like a pointed, bumpy green cucumber. It has a long history of use for diabetes in some Asian countries. Extracts of the fruit are now commonly used in North America. Bitter melon is thought to work through an insulin-like action. It contains an insulin-like polypeptide known as "plant insulin," "P-insulin," or "polypeptide P."

Bitter melon

Several preliminary studies have evaluated bitter melon fruit, fruit juice, and fruit extract.[3-6] According to this preliminary research, bitter melon appears to lower blood glucose and A1C in patients with type 2 diabetes. However, available studies were small, some were short-term, and some were poorly designed. More high-quality research is needed to determine the significance of these potential benefits.

The most appropriate dose or dosage form is not known.

Safety Considerations/Drug Interactions

No adverse effects have been reported in studies evaluating bitter melon. Bitter melon contains an insulin-like constituent. Combining bitter melon with other hypoglycemic drugs such as glyburide might increase the risk of hypoglycemia.[7]

Chromium

Chromium was first discovered in France in the late 1790s. It took until the 1960s to discover its role in insulin function. In the 1970s, interest in using chromium for diabetes patients grew when a patient on long-term total parenteral nutrition developed symptoms of diabetes. It was discovered that these patients had a chromium deficiency. When chromium was given, diabetes symptoms resolved.[8] It is now known that chromium plays a key role in insulin function. Chromium is the active component of glucose tolerance factor, which is a complex of chromium bound to single molecules of glycine, glutamic acid, cysteine, and two molecules of nicotinic acid.[8] Chromium deficiency is rare, but it is thought that some diabetes patients excrete more chromium and therefore are at increased risk for deficiency.[9]

Several, but not all, clinical studies suggest that chromium can be beneficial for some diabetes patients.[10] In a controlled trial of 78 elderly patients taking 200 mcg of chromium twice daily for 3 weeks, there was a significant difference in the fasting blood level of glucose compared to the baseline (190 mg/dL versus 150 mg/dL, $p < 0.001$) and significant A1C improvement from 8.2% to 7.6% ($p < 0.01$).[11] Chromium might also have the added benefit of helping to lower triglyceride and cholesterol levels in diabetes patients.[10] This suggests a potential role for chromium in treating patients with metabolic syndrome. But there are still questions about how effective chromium is and for which patients it might be most beneficial. About 40% to 80% of diabetes patients who take chromium supplements have improved blood glucose.[11] There is some speculation that only diabetes patients with a chromium deficiency might benefit from taking chromium.

Most studies involving chromium for diabetes have used chromium picolinate. Doses have ranged from 200 to 1000 mcg/day taken in divided doses. It is not known if higher doses are more effective. But taking a higher dose does seem to result in a quicker response.

Safety Considerations/Drug Interactions

Chromium is usually well tolerated, but some patients experience side effects, even at low doses. These can include cognitive or perceptual dysfunction, headache, insomnia, irritability, and mood changes.[8] There is some concern that chromium might adversely affect psychiatric disorders. Chromium is usually taken as chromium picolinate. The picolinate component can alter levels of serotonin, norepinephrine, and dopamine.[8] Theoretically, this might lead to psychiatric complications. This effect has not been documented in the literature.

Nonsteroidal anti-inflammatory drugs (NSAIDs) increase absorption and retention of chromium.[12] Theoretically, this could lead to excessive chromium levels and potentially chromium toxicity. Chromium in higher doses than discussed here can cause stomach upset, ulcers, convulsions, and kidney and liver damage.

CASSIA CINNAMON

Cinnamon refers to several different varieties of the spice. Cinnamomum zeylanicum, also known as Ceylon cinnamon, is the type most common in the Western world. Cinnamomum aromaticum, also known as cassia cinnamon or Chinese cinnamon, is also commonly used. Cinnamon has been used for centuries as a medicinal spice. But it has only been recently that interest in using cinnamon for diabetes grew. Polyphenols in cinnamon such as hydroxychalcone are thought to phosphorylate the insulin receptor. This increases insulin sensitivity and increases tissue uptake of glucose.[13,14]

PATIENT CASE

KEVIN

Kevin is a 65-year-old male with newly diagnosed type 2 diabetes. He is 5'10", with a weight of 208 pounds and a waist circumference of 40 inches. Kevin is retired from his job as an accountant. He does not exercise. He lives alone and consumes cereal for breakfast. He eats primarily fast food, with occasional restaurant meals, for lunch and dinner. He was also found to have elevated blood pressure at the time of diagnosis. His physician started him on lisinopril 10 mg/day for hypertension. He was titrated to glyburide 10 mg twice daily to control his blood glucose levels.

Kevin received a blood glucose monitor and has been checking his levels at home. He reports that his FPG in the morning averages 147 mg/dL (range 113 to 165) and his PPG after lunch averages 223 mg/dL (range 198 to 250), over the past 3 months. On his most recent visit to his physician, his A1C was 8.2%. His physician wanted to prescribe another drug to help control his blood glucose, but Kevin declined. He explains that he would rather use dietary and other natural approaches. He asks for advice about what natural products he should take to reduce his blood glucose and A1C levels.

In 2003, the first clinical research was published suggesting cassia cinnamon improves blood glucose. The evidence is preliminary, but promising. Patients with type 2 diabetes who took 1 to 6 g/day of cinnamon for 40 days had fasting blood glucose levels lowered by 18% to 29%. Cinnamon also seems to lower triglycerides by 23% to 30%, low-density lipoprotein (LDL) cholesterol by 7% to 27%, and total cholesterol by 12% to 26%.[15]

The only clinical evidence to date involves cassia cinnamon. It is not known if other varieties of cinnamon are effective. A dose of 1 to 6 g was used in a clinical trial. One teaspoon of cinnamon is equivalent to 4.75 g.

SAFETY CONSIDERATIONS/DRUG INTERACTIONS

No side effects have been reported and there are no known drug interactions.

GYMNEMA

Gymnema (Gymnema sylvestre) is a popular Indian plant. In Hindi, it is called "gurmar," which means "sugar destroying." It is thought to work in diabetes by increasing pancreatic beta-cell growth or by stimulating beta-cells to release more insulin.[16,17]

Two preliminary clinical studies have evaluated a specific gymnema extract (GS4) 400 mg/day in patients with type 1 and 2 diabetes. In both studies, the extract significantly reduced fasting blood glucose levels and A1C; however, the studies were not controlled and involved small numbers of patients.[18,19] More information is needed to verify the effectiveness of gymnema.

SAFETY CONSIDERATIONS/DRUG INTERACTIONS

Gymnema extract has been safely used in small studies lasting up to 20 months. No adverse effects have been reported. A possible hypoglycemic effect should be considered.

PSYLLIUM

When people talk about psyllium, they are usually referring to "blond psyllium" (Plantago ovata). This is the kind of psyllium in most products such as Metamucil®.

Fiber has beneficial effects besides treating constipation. It can reduce postprandial glucose levels in patients with diabetes by slowing intestinal transit. It also binds fats and has a beneficial effect on the lipid profile.

Blond psyllium significantly reduces postprandial glucose levels by 14% to 20% and insulin levels in type 2 diabetes patients according to numerous randomized clinical trials.[20-25] Doses of 15 g/day in divided doses were used in the clinical studies. Psyllium should be taken with meals.

SAFETY CONSIDERATIONS/DRUG INTERACTIONS

Psyllium is very safe when used appropriately. It is usually well tolerated, but it can cause transient abdominal bloating and pain, flatulence, and nausea. Starting with low doses and titrating up can help minimize these side effects.[8]

Some patients are allergic to psyllium. Allergic reactions in these patients can range from a minor skin rash to life-threatening anaphylaxis.[26-30] However, allergic reactions are uncommon. Psyllium should be avoided in patients with certain gastrointestinal conditions such as fecal impaction or obstruction or swallowing disorders. Psyllium may bind to oral medications. These should be administered separately.

STEVIA

Stevia rebaudiana is a native plant in Paraguay. The leaves of stevia have traditionally been used as a sweetener. In the United States, stevia is marketed as a dietary supplement and is often promoted for diabetes. Two constituents in stevia called steviol and stevioside might increase insulin release from pancreatic beta-cells. There is also some evidence that these stevia constituents might increase tissue insulin sensitivity or decrease glucose absorption in the gut.[31-33]

Clinical evidence supporting stevia for diabetes is scant. One small clinical trial suggests a stevia extract can decrease postprandial glucose levels by about 18% in patients with type 2 diabetes.[34] The long-term effects of stevia on diabetes are unknown. A product providing 1 g of the stevioside constituent was used in this study.

SAFETY CONSIDERATIONS/DRUG INTERACTIONS

The FDA does not allow stevia to be sold on the market as a sweetener because there is not enough evidence about its safety. Dietary supplements containing 1 to 1.5 g of stevioside constituent seem to be safe when used for up to 2 years. The most commonly reported side effects include nausea, feeling of fullness, headache, dizziness, and myalgia.[8] Stevia is a member of the Asteraceae/Compositae plant family, which includes ragweed, marigolds, daisies, and others. Patients allergic to these plants should avoid stevia. There are no known drug interactions.

PRICKLY PEAR CACTUS

Prickly pear cactus (Opuntia species) is a plant remedy that is used primarily in the Mexican and Mexican-American cultures as a treatment for type 2 diabetes and as a food product. The broiled stems of the species Opuntia streptacantha have demonstrated the ability to decrease blood glucose.[35] The

Prickly pear cactus

effect is attributable to the high fibrous polysaccharide content, including pectin, which can slow carbohydrate absorption from the gut and also possibly due to an insulin-sensitizing effect.[8]

Prickly pear cactus administered as a single dose decreases blood glucose by 17% to 46% in some patients.[35-38] These studies used broiled cactus stems.[36-38] The typical doses range from 100 to 500 g/day, divided three times daily.[8] This dose is not practical. It would require eating a large serving of opuntia with each meal. Patients who use this product are usually from Mexico and take it "as needed" for elevated blood glucose. This method does not work well either. Opuntia has been studied as a dried extract, which would be much more feasible; however, this was found to be ineffective.[36]

SAFETY CONSIDERATIONS/DRUG INTERACTIONS

Prickly pear cactus is a safe natural product when used orally. It is usually well tolerated; however, it can cause headache and some gastrointestinal side effects that include mild diarrhea, nausea, increased stool volume, increase stool frequency, and abdominal fullness.[8] The gastrointestinal effects are expected when beginning consumption of bulk-forming agents.

This product may have an internal mechanism as well and could pose a risk for hypoglycemia, particularly when taken with hypoglycemic drugs. Patients who use prickly pear cactus with other hypoglycemic agents should be cautioned on hypoglycemic symptoms and management of these symptoms.

AMERICAN GINSENG

The applicable part of American ginseng (Panax quinquefolius) is the root. It matures slowly, making harvest available only every few years. It was once abundant in parts of eastern North America, but it is now considered a rare or endangered species in many regions because of overharvesting.[39] American ginseng is used as an adaptogen, for increasing the body's resistance to biological and environmental stresses.

The effects of American ginseng on blood glucose seem to be related to the constituents ginsenosides, nonsaponin peptidoglycans, and possibly other nongensinoside constituents. The actions of these constituents may be responsible for the primary effect, reduction of postprandial glucose levels.[40,41] American ginseng also contains nonsaponin peptidoglycans called quinquefolans, which have hypoglycemic effects.[42]

A small clinical trial to assess reduction in postprandial glucose with dose escalation and time of American ginseng administration in type 2 diabetic patients demonstrated that significant reductions in "incremental glycemia" of 13% to 59% were achieved with 3, 6, and 9 g of American ginseng at 30, 45, and 120 minutes before an oral glucose challenge.[41]

SAFETY CONSIDERATIONS/DRUG INTERACTIONS

In general, the use of American ginseng has not been associated with serious adverse reactions.[43] Diarrhea and allergic skin reactions can occur with related species of Panax ginseng, especially with large amounts or prolonged use.[8] American ginseng should not be taken during pregnancy. Ginosenside Rb1, an active constituent of this natural product, has teratogenic effects in animal models.[44]

American ginseng has been shown to lower blood glucose; therefore, use caution with other hypoglycemics. The concomitant use of American ginseng may potentiate the activity of stimulant drugs, including caffeine.[45,46] It can also decrease the effectiveness of warfarin therapy.[47,48] Other possible interactions include antipsychotic drugs and monoamine oxidase (MAO) inhibitors.[8] The dose to reduce postprandial glucose levels in type 2 diabetes patients is 3 to 9 g up to 2 hours before a meal.[40,41]

PATIENT DISCUSSION

None of the products discussed in this chapter has been studied enough to determine with certainty the efficacy regarding lowering blood glucose. The best choice of product to try is one that has some degree of evidence, and importantly, a good safety record. Cassia cinnamon fits the criteria and a recommendation could be made to Kevin to sprinkle a gram of cassia cinnamon on his cereal in the morning. Psyllium is also very safe and can be used to help decrease his postprandial levels. It will also help improve his lipid profile. Chromium might be worth a try. Kevin can see if his physician will order a chromium level. Chromium appears to offer more improvement in patients with a known deficiency. If not, he can give chromium supplementation (at safe levels) a trial period to see if his glucose levels improve.

Prickly pear cactus can be tried, if the patient likes it. The taste is somewhat tart. Patients of Mexican descent often blend ground cactus with sweet juices, counteracting any potential benefit. The cactus should be prepared in a healthy manner, such as broiling. Not all prickly pear cactus has demonstrated benefit, and different species are available. Wild American ginseng is an endangered species in the United States and should not be purchased. Cultivated American ginseng is available, if Kevin wishes to try this product. The other products need more data before their use can be recommended.

He should be counseled to try one product at a time, in a recommended dose and for a reasonable trial period. He should also pursue the lifestyle changes described earlier. If he cannot get his blood glucose under control quickly, he should be encouraged to add on a prescription medication.

SELF-ASSESSMENT

1. What is TRUE concerning bitter melon:
   a. It has a long history of use for diabetes in some Asian countries.
   b. It has an insulin-like action that lowers blood glucose in patients with type 2 diabetes.
   c. There is an absence of adverse effects reported in studies evaluating bitter melon.
   d. All of the above.

2. Chromium may have benefit in lowering triglycerides and cholesterol in addition to decreasing blood glucose and A1C.
   a. True
   b. False

3. Which one of the following drugs could increase chromium levels?
   a. ACE inhibitors
   b. Beta adrenergic blockers
   c. HMG-CoA inhibitors
   d. NSAIDs

4. Which is FALSE concerning cassia cinnamon:
   a. Phosphorylation of the insulin receptor by polyphenols found in cassia cinnamon increases insulin sensitivity.
   b. Cassia cinnamon is known to have significant drug interactions.
   c. There is a lack of reported side effects and drug interactions with the use of cinnamon.
   d. Cassia cinnamon doses of 1 to 6 g daily were used in a clinical trial.

5. Preliminary studies demonstrate that gymnema extract reduces fasting blood glucose levels and A1C.
   a. True
   b. False

6. What is the dose of gymnema (specific extract GS4) used for diabetes in clinical trials?
   a. 4 mg/day
   b. 40 mg/day
   c. 400 mg/day
   d. 400 mg twice daily

7. Which is TRUE concerning psyllium:
   a. Psyllium is used in patients with diabetes to lower postprandial glucose levels.
   b. Psyllium causes gastrointestinal side effects.
   c. Patients using psyllium must ensure adequate fluid intake.
   d. All of the above.

8. Stevia is used as a sweetener and for lowering blood glucose.
   a. True
   b. False

9. Which is FALSE concerning prickly pear cactus:
   a. The broiled stems of Opuntia streptacantha may decrease postprandial glucose levels.
   b. Prickly pear cactus should be prepared in a healthy manner.
   c. Prickly pear cactus can be taken in supplement form.
   d. Prickly pear cactus is usually well tolerated.

10. American ginseng can lower the effect of this drug:
    a. Warfarin
    b. Atenolol
    c. Hydrochlorothiazide
    d. Amiodarone

Answers: 1-d; 2-a; 3-d; 4-b; 5-a; 6-c; 7-d; 8-a; 9-c; 10-a

REFERENCES

1. Standards of Medical Care in Diabetes. V. Diabetes Care. *Diabetes Care.* 2005;28 (Suppl 1):S8-14.

2. Pizzorno JE, Murray MT, Joiner-Bey H. *The Clinician's Handbook of Natural Medicine.* London: Churchill Livingstone; 2003.

3. Welihinda J, Karunanayake EH, Sheriff MH, et al. Effect of Momordica charantia on the glucose tolerance in maturity onset diabetes. *J Ethnopharmacol.* 1986;17:277-82.

4. Srivastava Y, Venkatakrishna-Bhatt H, Verma Y, et al. Antidiabetic and adaptogenic properties of Momordica charantia extract: an experimental and clinical evaluation. *Phytother Res.* 1993;7:285-9.

5. Baldwa VS, Bhandari CM, Pangaria A, et al. Clinical trial in patients with diabetes mellitus of an insulin-like compound obtained from plant sources. *Upsala J Med Sci.* 1977;82:39-41.

6. Ahmad N, Hassan MR, Halder H, et al. Effect of Momordica charantia (Karolla) extracts on fasting and postprandial serum glucose levels in NIDDM patients. *Bangladesh Med Res Counc Bull.* 1999;25:11-3. Abstract.

7. Aslam M, Stockley IH. Interaction between curry ingredient (karela) and drug (chlorpropamide). *Lancet.* 1979:1:607.

8. Jellin JM, Gregory PJ, Batz F, et al. Natural Medicines Comprehensive Database. Therapeutic Research Faculty; Stockton CA; 2004. Available at www.naturaldatabase.com. Accessed December 27, 2004.

9. Liu VJ, Abernathy RP. Chromium and insulin in young subjects with normal glucose tolerance. *Am J Clin Nutr.* 1982;35:661-7.

10. Anderson RA, Cheng N, Bryden NA, et al. Elevated intakes of supplemental chromium improve glucose and insulin variables in individuals with type 2 diabetes. *Diabetes.* 1997;46:1786-91.

11. Althius MD, Jordon NE, Ludington EA, et al. Glucose and insulin responses to dietary chromium supplements: a meta-analysis. *Am J Clin Nutr.* 2002;76:148-55.

12. Food and Nutrition Board, Institute of Medicine. Dietary Reference Intakes for Vitamin A, Vitamin K, Arsenic, Boron, Chromium, Copper, Iodine, Iron, Manganese, Molybdenum, Nickel, Silicon, Vanadium, and Zinc. Washington, DC: National Academies Press; 2000. Available at www.nap.edu/books/0309072794/html.

13. Anderson RA, Broadhurst CL, Polansky MM, et al. Isolation and characterization of polyphenol type-A polymers from cinnamon with insulin-like biological activity. *J Agric Food Chem.* 2004;52:65-70.

14. Jarvill-Taylor KJ, Anderson RA, Graves DJ. A hydroxychalcone derived from cinnamon functions as a mimetic for insulin in 3T3-L1 adipocytes. *J Am Coll Nutr.* 2001;20:327-36.

15. Khan A, Safdar M, Ali Khan M, et al. Cinnamon improves glucose and lipids of people with type 2 diabetes. *Diabetes Care.* 2003;26:3215-8.

16. Persaud SJ, Al-Majed H, Raman A, et al. Gymnema sylvestre stimulates insulin release in vitro by increased membrane permeability. *J Endocrinol.* 1999;163:207-12.

17. Sinsheimer JE, Rao GS, McIlhenny HM. Constituents from Gymnema sylvestre leaves. V. Isolation and preliminary characterization of the gymnemic acids. *J Pharm Sci.* 1970;59:622-8.

18. Shanmugasundaram ER, Rajeswari G, Baskaran K, et al. Use of Gymnema sylvestre leaf extract in the control of blood glucose in insulin-dependent diabetes mellitus. *J Ethnopharmacol.* 1990;30:281-94.

19. Baskaran K, Kizar-Ahamath B, Shanmugasundaram MR, et al. Antidiabetic effect of leaf extract from Gymnema sylvestre in non-insulin-dependent diabetes mellitus patients. *J Ethnopharmacol.* 1990;30:295-300.

20. Anderson JW, Allgood LD, Turner J, et al. Effects of psyllium on glucose and serum lipid responses in men with type 2 diabetes and hypercholesterolemia. *Am J Clin Nutr.* 1999;70:466-73.

21. Pastors JG, Blaisdell PW, Balm TK, et al. Psyllium fiber reduces rise in postprandial glucose and insulin concentrations in patients with non-insulin-dependent diabetes. *Am J Clin Nutr.* 1991;53:1431-5.

22. Wolever TM, Vuksan V, Eshuis H, et al. Effect of method of administration of psyllium on glycemic response and carbohydrate digestibility. *J Am Coll Nutr.* 1991;10:364-71.

23. Frati Munari AC, Benitez Pinto W, Raul Ariza Andraca C, et al. Lowering glycemic index of food by acarbose and Plantago psyllium mucilage. *Arch Med Res.* 1998 Summer;29:137-41.

24. Rodriguez-Moran M, Guerrero-Romero F, Lazcano-Burciaga G. Lipid- and glucose-lowering efficacy of Plantago psyllium in type II diabetes. *J Diabetes Complications.* 1998;12:273-8.

25. Sierra M, Garcia JJ, Fernandez N, et al. Therapeutic effects of psyllium in type 2 diabetic patients. *Eur J Clin Nutr.* 2002;56:830-42.

26. Suhonen R, Kantola I, Bjorksten F. Anaphylactic shock due to ingestion of psyllium laxative. *Allergy.* 1983;38:363-5.

27. Vaswani SK, Hamilton RG, Valentine MD, et al. Psyllium laxative-induced anaphylaxis, asthma, and rhinitis. *Allergy.* 1996;51:266-8.

28. Freeman GL. Psyllium hypersensitivity. *Ann Allergy.* 1994;73:490-2.

29. Lantner RR, Espiritu BR, Zumerchik P, et al. Anaphylaxis following ingestion of a psyllium-containing cereal. *JAMA.* 1990;264:2534-6.

30. Kaplan MJ. Anaphylactic reaction to "Heartwise." *N Engl J Med.* 1990;323:1072-3.

31. Jeppesen PB, Gregersen S, Poulsen CR, et al. Stevioside acts directly on pancreatic beta cells to secrete insulin: actions independent of cyclic adenosine monophosphate and adenosine triphosphate-sensitive K+-channel activity. *Metabolism.* 2000;49:208-14.

32. Lailerd N, Saengsirisuwan V, Sloniger JA, et al. Effects of stevioside on glucose transport activity in insulin-sensitive and insulin-resistant rat skeletal muscle. *Metabolism.* 2004;53:101-7.

33. Toskulkao C, Sutheerawatananon M, Wanichanon C, et al. Effects of stevioside and steviol on intestinal glucose absorption in hamsters. *J Nutr Sci Vitaminol* (Tokyo). 1995;41:105-13.

34. Gregersen S, Jeppesen PB, Holst JJ, Hermansen K. Antihyperglycemic effects of stevioside in type 2 diabetic subjects. *Metabolism.* 2004;53:73-6.

35. Frati AC, Xilotl Diaz N, Altamirano P, et al. The effect of two sequential doses of *Opuntia streptacantha* upon glycemia. *Arch Invest Med.* (Mex) 1991;22(3-4):333-6.

36. Frati-Munari AC, Altamirano-Bustamante E, Rodrigues-Barcenas N, et al. Hypoglycemic action of *Opuntia streptacantha* Lemaire: study using raw extracts. *Arch Invest Med.* (Mex) 1989;20:321-5.

37. Frati-Munari AC, Del Valle-Martinez LM, Ariza-Andraca CR, et al. Hypoglycemic action of different doses of nopol (*Opuntia streptacantha*) in patients with type II diabetes mellitus. *Arch Invest Med.* (Mex) 1989;20:197-201.

38. Frati-Munari AC, Gordillo BE, Altamirano P, et al. Hypoglycemic effect of *Opuntia streptacantha* Lemaire in NIDDM. *Diabetes Care.* 1988;11:63-6.

39. Peirce A. The American Pharmaceutical Association Practical Guide to Natural Medicines. New York: The Stonesong Press. 1999:296-8.

40. Vuksan V, Sievenpiper JL, Koo VY, et al. American ginseng (*Panax quinquefolius* L) reduces postprandial glycemia in nondiabetic subjects and subjects with type 2 diabetes mellitus. *Arch Intern Med.* 2000;160:1009-13.

41. Vuksan V, Stavro MP, Sievenpiper JL, et al. Similar postprandial glycemic reductions with escalation of dose and administration time of American ginseng in type 2 diabetes. *Diabetes Care.* 2000;23:1221-6.

42. Sievenpiper JL, Arnason JT, Leiter LA, et al. Decreasing, null and increasing effects of eight popular types of ginseng on acute postprandial glycemic indices in healthy humans: the role of ginsenosides. *J Am Coll Nutr.* 2004;23:248-58.

43. McElhaney JE, Gravenstein S, Cole SK, et al. A placebo-controlled trial of a proprietary extract of North American ginseng (CVT-E002) to prevent acute respiratory illness in institutionalized older adults. *J Am Geriatr Soc.* 2004;52:13-9.

44. Chan LY, Chiu PY, Lau TK. An in-vitro study of ginsenoside Rb(1)-induced teratogenicity using a whole rat embryo culture model. *Hum Reprod.* 2003;18:2166-8.

45. Newall CA, Anderson LA, Philpson JD. *Herbal Medicine: A Guide for Healthcare Professionals.* London: The Pharmaceutical Press; 1996.

46. McGuffin M, Hobbs C, Upton R, Goldberg A, eds. *American Herbal Products Association's Botanical Safety Handbook.* Boca Raton, FL: CRC Press, LLC; 1997.

47. Janetzky K, Morreale AP. Probable interaction between warfarin and ginseng. *Am J Health Syst Pharm.* 1997;54:692-3.

48. Yuan CS, Wei G, Dey L, et al. American ginseng reduces warfarin's effect in healthy patients: a randomized, controlled trial. *Ann Intern Med.* 2004;141:23-7.

Chapter 13
Cold and Flu

Karen Shapiro

Millions of people develop the common cold every year and it results in lots of missed work and plenty of misery. When a reliable remedy to prevent or treat the common cold comes out, it will undoubtedly be the best-selling medicine ever.

That remedy is not here yet. There is some evidence supporting almost all of the remedies discussed in this chapter. But, even if these remedies do help some patients have some relief from their symptoms, they are not a magic bullet. Benefits are modest at best.

The products discussed here for influenza prevention and treatment are based on studies of the normal seasonal respiratory illness caused by influenza. Pandemic viruses can emerge as a result of "antigenic shift" which causes a sudden, major change in the virus. The efficacy for prevention and treatment of these more serious influenza virus subtypes using these agents is not known.

KEY POINTS

- There is no magic bullet natural product for preventing or treating the common cold or flu.
- Andrographis appears to reduce the symptoms and duration of a cold and may be useful for prevention.
- Echinacea may provide modest benefit for cold treatment; data at present are contradictory.
- Preliminary data suggest that elderberry is useful for influenza treatment.
- Vitamin C may decrease the duration of a cold; data is contradictory.
- Echinacea, vitamin C, and zinc do not appear useful in cold prevention.
- Intranasal zinc has been linked to anosmia (loss of smell) in some patients that could be permanent.
- There is preliminary evidence that Asian and American ginseng may provide modest benefit in influenza prevention.
- Oscillococcinum is a homeopathic product that contains no detectable active ingredient and likely has no medicinal value, except for a possible placebo effect.

Product	Dosage	Effect	Safety Concerns
Andrographis	400 mg three times daily	Several trials including > 800 patients; significant reduction in cold symptoms and duration	• Urticaria, gastrointestinal upset
Echinacea	No typical dose	Possible decrease in cold symptom duration; contradictory evidence; no benefit for use as a prophylactic agent	• Allergic reactions
Vitamin C	1 to 3 g/day	Multiple clinical trials suggest modest decrease in cold symptom duration; some contradictory evidence; no benefit for use as a prophylactic agent	• Gastrointestinal upset, osmotic diarrhea with high doses • May affect levels of certain drugs; see text
Zinc	Lozenges: One lozenge providing 9 to 24 mg of zinc every 2 hours. Intranasal: One inhalation per nostril every 4 hours.	Multiple clinical trials suggest decrease in cold symptom duration; some contradictory evidence; no benefit for use as a prophylactic agent	• Metallic tastes • Copper deficiency • Anosmia with intranasal zinc
Elderberry	1 tablespoon syrup every 4 hours for 3 to 5 days	Two clinical trials showing surprising reduction in duration of flu symptoms	• Nausea, gastrointestinal upset
Ginseng	Asian ginseng extract: 100 mg daily. American ginseng extract: 200 mg twice daily	Two preliminary clinical trials for preventing flu	• Insomnia, headache, palpitations, tachycardia • Concern for patients with cardiovascular conditions
Oscillococcinum	Contents of one tube up to three times daily	Several poor quality trials; suggestion of possible reduction in flu symptom duration	• None known or expected

ANDROGRAPHIS

Andrographis (*Andrographis paniculata*) is a traditional Indian (Ayurvedic) medicine. Historically, it has been used as a "blood purifier" and as an antipyretic. Today, its most common use is as an immunostimulant for increasing resistance to viral infections such as the common cold.[1]

The leaf and rhizome of the andrographis plant are used to make the medicinal extract. The active constituents of the extract are thought to be primarily andrographolide and possibly other constituents.[2,3] The extract seems to have an immunostimulant effect and increases antibody activity and phagocytosis by macrophages.

Several clinical trials involving more than 800 patients suggest that taking andrographis extract at the start of a cold significantly reduces symptoms such as sore throat, fatigue and tiredness, excessive nasal secretions, and earache.[4-9] In some cases, symptoms such as tiredness and sleeplessness improved within 2 days of starting andrographis. Other symptoms improved or resolved within 4 to 5 days.

In addition to treating cold symptoms, there is preliminary evidence that taking andrographis extract for 2 months might decrease the risk of getting a cold by about 50%. More evidence is needed to confirm this finding.[10]

Most clinical studies have used a specific andrographis extract standardized to contain 4 to 5.6 mg of andrographolide per tablet (Kan Jang, Swedish Herbal Institute). In clinical studies using this extract, a dose of 400 mg three times daily was used.

PATIENT CASES

JAMES—A PATIENT WITH A COLD

James is a 42-year-old male. He has had a cold for the past 2 days, with runny nose, sore throat, mild headache, and general feeling of malaise and fatigue. He asks if echinacea works, or if you can recommend something else. James states that he has no health problems. He has been to the physician recently and was told that he is in "excellent" health.

MAX—A PATIENT LOOKING FOR INFLUENZA PROTECTION AND TREATMENT

Max is a 54-year-old male who is concerned about getting the flu. He says he has diabetes and was told to get a flu shot each year. He went to get his shot late this year and found out that there is no more vaccine due to a shortage. He heard that there are some natural products that he can take. He asks for your advice about which products to use to keep him from getting the flu, and what to take if he gets it anyway.

SAFETY CONSIDERATIONS/DRUG INTERACTIONS

Andrographis extract has been well tolerated in clinical trials. Urticaria has been reported in some patients.[7] Excessive doses of andrographis cause gastrointestinal upset, anorexia, and vomiting. Some constituents of andrographis seem to have a hypotensive effect.[11] Theoretically, taking andrographis might exacerbate hypotension in a patient with low blood pressure. However, this effect has not been demonstrated in humans.

There are no documented drug interactions with andrographis. However, since andrographis has immunostimulant properties, there is some concern that it could decrease the effectiveness of immunosuppressants (e.g., cyclosporine and tacrolimus). Decreasing the effectiveness of the immunostimulant could result in organ rejection.

Some constituents of andrographis seem to have antiplatelet and antithrombotic effects.[12,13] Theoretically, combining andrographis with antiplatelet or anticoagulant drugs (e.g., aspirin or warfarin) might increase the risk of bruising and bleeding. This interaction has not been documented.

ECHINACEA

Echinacea has a long history of use in the United States. It was a common Native American remedy for infectious diseases and was eventually adopted by U.S. settlers. In the early 1900s, it was one of the most widely prescribed native botanicals. It began to fall out of favor as antibiotics were discovered and developed.

Echinacea is the most commonly used herbal product in the United States. In 2003, sales reached almost $40 million.[14]

Despite its popularity and long history of use, there is not a lot of strong evidence to support echinacea for treating the common cold. There have been several clinical studies involving more than 3000 patients with the common cold. Most of these suggest that taking echinacea improves symptoms of the common cold and decreases duration of symptoms by about 1 to 3 days.[15-20] Due

Echinacea

to methodological flaws, lack of consistency in study design, and the variety of echinacea formulations used in these studies, reliably quantifying echinacea's potential benefits for the common cold is difficult. Furthermore, some studies have shown no benefit from taking echinacea.[21-23] Findings from systematic reviews and meta-analyses suggest that, although echinacea might be more effective than placebo for treating the common cold, research to date is incomplete and inconclusive.[24-26]

For preventing the common cold, the evidence has been more consistent. Several clinical trials have found that taking echinacea prophylactically does not reduce the chance of developing a cold.[21,24,27,28]

There are actually three *Echinacea* species of medical interest. These include *E. angustifolia*, *E. pallida*, and *E. purpurea*. Some products contain just one of these species or a combination of them.

Clinical studies involving echinacea have not been consistent with regard to the type of Echinacea species used. Studies have used different species, different portions of the plant (e.g., root, leaves, whole plant), different types of extracts, and different dosage forms and doses. Therefore, there is no consensus regarding which is the most effective or most appropriate formulation to use.

SAFETY CONSIDERATIONS/DRUG INTERACTIONS

Echinacea products are usually very well tolerated and usually cause few side effects. Some side effects that have been reported include allergic reactions, fever, heartburn, diarrhea, constipation, dizziness, sore throat, dry mouth, and insomnia.[23,28,29]

Allergic reactions are the biggest concern in patients who use echinacea. In a study in children, as many as 7% of patients who took echinacea experienced a rash that may have been the result of an allergic reaction.[30] Allergic reactions in adults and children who have taken echinacea range from minor skin rash to angioedema and anaphylaxis.[31-33] Some people seem to be more susceptible to allergic reactions to echinacea. People who are allergic to ragweed, daisies, marigolds, and other plants in the Asteraceae/Compositae family are more likely to be sensitive to echinacea, which is in the same plant family.

Patients with atopy, a genetic disposition to allergies, might also be more likely to be sensitive to echinacea. In one study, 20% of 100 atopic patients tested positive for hypersensitivity to echinacea using a skin test.[32]

Since echinacea seems to have immunostimulatory effects, there is some concern that it might exacerbate autoimmune disorders. There is one report of a patient developing Sjogren's syndrome after taking a herbal combination product containing echinacea, St. John's wort, and kava.[34] In another case, a patient with well-controlled pemphigus vulgaris experienced a significant disease flare within 1 week of taking echinacea. Symptoms improved when echinacea was discontinued.[35] Due to echinacea's immunostimulant properties, there is also concern that it could counteract or decrease the effectiveness of immunosuppressant drugs such as cyclosporine. The potential consequences of this theoretical interaction could include transplant rejection.

In vitro evidence suggests that *E. angustifolia* root extract inhibits cytochrome P450 3A4 (CYP 450 3A4).[36] Theoretically, taking echinacea with CYP 450 3A4 substrates might increase drug levels and potentially increase the risk of adverse effects. This interaction has only been documented for midazolam (Versed®). Taking echinacea and midazolam concurrently significantly increases levels of midazolam.[37]

VITAMIN C

Vitamin C, also known as ascorbic acid, is one of the most commonly used vitamins. It is a water-soluble vitamin that is available in high concentrations in many vegetables and fruits, particularly citrus fruits such as oranges.

Vitamin C was first made famous in 1747 when a Scottish naval surgeon discovered that a nutrient in citrus fruits prevented scurvy. Vitamin C was also the first vitamin to be synthesized in a laboratory in 1935. More recently, two-time Nobel Prize winner Linus Pauling helped popularize vitamin C by promoting gram-sized doses for preventing many diseases.

Even though people have been taking whopping doses of vitamin C for years, the evidence supporting such practice has been scant. For the common cold, there have been several studies evaluating the effectiveness of taking large doses of vitamin C. There have been some conflicting findings, but the majority of evidence suggests taking 1 to 3 g of vitamin C daily might decrease the duration of cold symptoms by about 1 to 1.5 days.[38-42] Taking vitamin C prophylactically does not seem to reduce the risk of developing a cold.[36,38,39]

Studies showing benefit for the common cold have used doses of vitamin C from 1 to 3 g day. Doses of 2 g/day might be more effective than 1 g/day.[38,42] But there is no reliable evidence that doses greater than 3 g are any more effective.

Vitamin C products marketed as "Ester-C" claim to be more effective than other vitamin C products. Ester-C products typically contain calcium ascorbate along with small amounts of vitamin C metabolites. There is no reliable evidence that Ester-C products are more bioavailable or more effective than other vitamin C products.

Some manufacturers of vitamin C claim that "natural" vitamin C is more effective or more bioavailable than "synthetic" vitamin C. There is no reliable evidence to support this claim.

Safety Considerations/Drug Interactions

Vitamin C is usually well tolerated, even if large doses are used. But the risk of side effects increases as doses get larger. Reported side effects include nausea, vomiting, diarrhea, cramping, heartburn, esophagitis, fatigue, headache, and others.[43,44]

The tolerable upper intake level of vitamin C set by the Institute of Medicine is 2 g/day. Beyond this dose, the risk of side effects is significant, particularly osmotic diarrhea and gastrointestinal upset.[43] Since high doses of vitamin C are often used for the common cold, many patients might not find the potential modest benefit worth the risk of significant gastrointestinal upset.

Vitamin C seems to increase absorption of aluminum.[45,46] For most patients, this is not likely to be a significant problem. In patients with renal insufficiency, this could potentially lead to aluminum accumulation and toxicity.

Vitamin C 1 g day seems to reduce serum levels of indinavir (Crixivan®) by about 14% through an unknown mechanism.[47] It is not known if higher doses of vitamin C would decrease levels further or if vitamin C might affect levels of other protease inhibitors.

Very high doses of vitamin C (> 5 g/day) seem to decrease the response to warfarin (Coumadin).[47-50] The higher the vitamin C dose, the more significant the interaction. Vitamin C most likely decreases warfarin absorption by causing diarrhea.

Zinc

Zinc is the second most abundant trace element in the body. Zinc is involved in many biological processes including DNA and protein synthesis, immune function, wound healing, reproduction, taste and smell, and others. Zinc is available in high concentrations in several foods including seafood, dairy products, whole grains, and nuts and legumes.

Most studies on zinc and the common cold have evaluated the use of zinc lozenges. Most of the evidence indicates that taking zinc lozenges immediately after symptom onset can significantly decrease the duration of cold symptoms.[51-55] But not all findings have been positive. Some studies show no benefit.[56-58] The reason for the different outcomes among studies is unclear. One explanation is that the differences are due to different zinc formulations. It is thought that zinc works against the cold virus only in the ionized form.

Lozenges that do not successfully release adequate amounts of ionized zinc might not be effective. Some formulations of zinc lozenges with flavoring agents added such as citric acid might not release adequate amounts of zinc ion and therefore may not be as effective.

Positive studies with zinc lozenges have also been criticized due to inconsistent methodologies and criteria for evaluating cold symptoms as well as poor blinding of the treatment. Meta-analyses of zinc lozenge trials suggests that research is currently inconclusive.[59,60] Some studies evaluating intranasal zinc preparations have shown decreased duration of symptoms, but other studies have found no benefit.[61-63] Neither the zinc lozenges nor nasal spray appears to be beneficial for prophylaxis against the common cold.[64-66]

Zinc gluconate and zinc acetate lozenges have been used in clinical studies. Lozenges are given every 2 hours while awake providing a dose of 9 to 24 mg of zinc per lozenge. Intranasal zinc gluconate has been used in clinical studies, one inhalation per nostril every 4 hours.

Safety Considerations/Drug Interactions

Zinc is usually well tolerated, but many patients experience a metallic taste. Some patients report nausea and vomiting. The tolerable upper intake level established by the Institute of Medicine is 40 mg/day. Long-term use of doses greater than this amount can cause copper deficiency and anemia.[67]

Intranasal zinc causes nasal irritation and bloody nose, bad taste, dry mouth, headache, and sore throat.[62,64] Some people experienced loss of smell (anosmia) following use of zinc gluconate nasal spray. In some cases, the loss of smell was permanent.[68-70] Zinc might adversely affect olfactory neurons.

Taking zinc orally can bind oral antibiotics in the quinolone and tetracycline classes. Potassium-sparing diuretics might decrease zinc elimination, increase zinc levels, and possibly cause zinc toxicity.

Elderberry

The previous items discussed are used for cold prevention and treatment. This agent is used for influenza. In the middle ages, elderberry (*Sambucus nigra*) was purportedly a "holy tree" which was capable of maintaining or restoring health and promoting a long life. Native Americans also had great respect for elderberry. The berries and leaves were used both as a food source and as a medicine. The branches were used to make arrow shafts.

Today, elderberry products are sold as dietary supplements and primarily promoted as a treatment for the flu. The flavonoids in elderberry are thought to be the active constituents. Some elderberry extracts are standardized based on flavonoid content. Elderberry seems to have both antiviral and immunomodulating effects. It seems to stimulate production of interleukins and tumor necrosis factor. Elderberry also seems to inhibit replication of influenza A and B.[71,72]

Elderberry

The clinical evidence for elderberry looks promising but is preliminary. Two clinical trials have evaluated a syrup containing a specific elderberry juice extract (Sambucol®, Nature's Way). Both studies found that elderberry offered surprising benefit for patients with the flu. In the first study, almost 90% of patients experienced complete symptom relief within 2 to 3 days of starting elderberry.[72,73] In the second study, patients who started elderberry within 48 hours of developing flu symptoms had relief an average of 4 days earlier compared to placebo.[73] There is no evidence regarding the use of elderberry for prophylaxis against influenza.

Only one formulation of elderberry, Sambucol, has been evaluated in clinical trials. Therefore, it is not known if other elderberry formulations would offer the same potential benefit. Sambucol is a elderberry juice extract in a syrup formulation. The product labeling suggests a dose of 2 teaspoons daily for adults. But the clinical studies evaluating Sambucol used a dose of 1 tablespoon (15 mL) four times daily for 3 to 5 days.[72,73]

Safety Considerations/Drug Interactions

Elderberry was very well tolerated when used in two clinical trials. No adverse effects were reported. But these trials were very short term, lasting 3 to 5 days. The safety of long-term use is not known.

Eating the raw fruit has been linked to nausea, vomiting, and severe diarrhea. Additional side effects including dizziness, numbness, and stupor have been reported following use of elderberry juice.[1]

There are no known interactions with conventional drugs. Since elderberry might have immunostimulant effects, it is theoretically possible that it could interfere with immunosuppressive therapy.

Ginseng

When people say "ginseng," they could actually be referring to one of three different varieties: American ginseng, Asian ginseng, or Siberian ginseng.

American ginseng (*Panax quinquefolius*) and Asian ginseng (*Panax ginseng*) are from the same genus, Panax. Panax is from the Greek word panakos, which translates to panacea. Although these two species are not entirely identical in chemical makeup, they are often considered interchangeable. Both contain the active constituents ginsenosides.

Asian ginseng is considered the "true" ginseng. Asian ginseng was in such high demand in China in the 1700s that American ginseng began to be imported and substituted. Siberian ginseng (*Eleutherococcus senticosus*) is from a different genus and does not have the same composition as American or Asian ginseng. However, Siberian ginseng is often used or substituted for other ginsengs because it is cheaper. Polysaccharides in American ginseng are thought to be responsible for effects on immune function. American ginseng seems to activate monocytes and natural killer cells and stimulate tumor necrosis factor, interleukin-2, and gamma-interferon.[74,75] Asian ginseng also seems to stimulate natural killer cell activity, increase antibody titers, and might have other immunostimulatory properties.

There is preliminary evidence that taking American or Asian ginseng might help prevent the flu. A preliminary study suggests that taking a specific Asian ginseng extract (Ginsana G115) before and after receiving an influenza vaccine seems to reduce the risk of developing influenza compared to people who just received the vaccination.[74] Another preliminary study shows that elderly patients in nursing homes who take a specific American ginseng extract (CVT-E002) seem to have a decreased risk of laboratory-confirmed influenza. But it does not seem to decrease the risk of developing symptoms, symptom severity, or duration.[76]

There is not enough evidence to determine the most appropriate formulation or dose of ginseng for preventing the flu. A specific American ginseng extract (CVT-E002) has been taken in a 200-mg formulation given twice daily over a period of 12 weeks.[76] In another Asian ginseng extract (Ginsana G115) study, the dose used was 100 mg/day for 4 weeks.[77]

Safety Considerations/Drug Interactions

American and Asian ginseng are usually well tolerated. Most side effects do not occur any more frequently than placebo.[78] One of the most commonly reported side effects among patients who take either American or Asian ginseng is insomnia.[77] Other side effects that have been reported include tachycardia, mastalgia, vaginal bleeding, decreased appetite, headache, palpitations, vertigo, and itchy skin.

Ginseng might also affect cardiac function. There is some evidence that Asian ginseng can prolong the QT interval and might also modestly decrease blood pressure.[79] The long-term effects of ginseng on cardiac function are unknown. American and Asian ginseng should be used cautiously or avoided in patients with cardiovascular conditions.

Both American and Asian ginseng seem to lower blood glucose levels.[80,81] Theoretically, combining ginseng with antidiabetes therapy such as insulin or hypoglycemic drugs might increase the risk of hypoglycemia. American and Asian ginseng also seem to reduce the effectiveness of warfarin (Coumadin®).[82,83]

Since ginseng seems to stimulate immune function, there is also concern that it might decrease the effectiveness of immunosuppressants such as cyclosporine. This interaction has not been documented. Ginseng might also interfere with monoamine oxidase inhibitors (MAOIs). There is a report of insomnia, headache, tremor, and hypomania in a patient who combined an unspecified ginseng species with phenelzine (Nardil®).[84,85]

OSCILLOCOCCINUM

Oscillococcinum® or "Oscillo" is a brand-name homeopathic remedy from the company Boiron. It is very popular in the United States for treating and preventing the flu. Boiron claims that it is the best-selling homeopathic flu remedy in the United States.

The mechanism of this and other homeopathic remedies is extremely controversial. Many conventional medical practitioners simply disregard homeopathic remedies because they are often so dilute that they contain no original active ingredient. Any true pharmacological effect cannot be explained by the laws of science as understood today.

Homeopathy's basic principles are that "like treats like" and "potentiation through dilution." In fact, homeopathic practitioners believe that the more dilute a substance is, the more potent it becomes. In homeopathy, influenza would be treated with an extreme dilution of a substance that normally causes influenza when taken in high doses. While investigating the Spanish flu in 1917, a French physician discovered a substance he called "oscillococci." He mistakenly thought that his oscillococci were the cause of the flu. The next step was to find an abundant source of these oscillococci and use it in a homeopathic dilution to create a remedy. He chose duck.[86] Oscillococcinum is a dilution of duck liver and heart extract. But it is so dilute that it literally contains none of the original "active" ingredient.

Unlike most homeopathic remedies, Oscillococcinum has some supporting research. Several studies have been completed, but most have severe methodological flaws. Based on a meta-analysis of this research, Oscillococcinum does not seem to be beneficial for preventing the flu.[84] The manufacturer of Oscillococcinum recommends taking the contents of one tube up to three times daily.

SAFETY CONSIDERATIONS/DRUG INTERACTIONS

Because Oscillococcinum contains no active ingredient, it is not expected to cause any beneficial or harmful effects or drug interactions.

PATIENT DISCUSSION

FOR TREATING A COLD

If James is interested in trying a natural product for his cold symptoms, any of the products discussed would be appropriate for short-term use. James should be discouraged from using any of these remedies long term because there is no evidence of benefit with long-term use and in some cases the safety of long-term use is unknown.

James should be discouraged from using zinc nasal spray due to concerns that it might cause anosmia. If he wants to use echinacea, he should be counseled about the risk of allergic reaction and questioned about whether he has a history of allergies.

In regard to prophylaxis, hand washing is more important than any of these products. To reduce his risk of catching colds in the future, he should be counseled to wash his hands after contact with a cold sufferer or with objects or surfaces that may be contaminated.

FOR PREVENTING OR TREATING THE FLU

For Max and all other persons with diabetes, there is no good alternative to the influenza vaccine. American and Asian ginseng are the only natural products with some evidence to support their use prophylactically against the flu, however the data is too preliminary to make this recommendation, especially in light of the known benefits from receiving the vaccine.

For treatment, the evidence for elderberry is promising, but too preliminary to recommend for most patients. In the event that Max develops the flu, elderberry might be worth a try. However, prescription antivirals are better studied. Oscillococcinum is likely to be harmless, but due to the lack of solid evidence and controversial nature of homeopathic remedies, it should not be recommended.

Self-Assessment

1. Which of the following statements concerning andrographis is TRUE:
   a. It may cause a reduction in sore throat, fatigue, and congestion.
   b. It may possess immunostimulating properties.
   c. There is a theoretical concern of increased bleeding risk.
   d. All of the above.

2. Which product can cause anosmia when used in a nasal formulation?
   a. Zinc
   b. Echinacea
   c. Elderberry
   d. Vitamin C

3. Which of the following *Echinacea* species is most effective for treating the common cold?
   a. E. angustifolia
   b. E. pallida
   c. E. purpurea
   d. This is unknown at present.

4. The most significant concern with the use of echinacea is:
   a. Halo vision
   b. Gastrointestinal upset
   c. Allergic reactions, including angioedema
   d. Headache

5. Linus Pauling advocated large doses of vitamin C for disease prevention. Doses greater than 2 g/day are likely to cause:
   a. Rash
   b. Diarrhea
   c. Constipation
   d. Peripheral neuropathy

6. "Natural" vitamin C is more effective than synthetic vitamin C for reducing the duration of a cold.
   a. True
   b. False

7. Oscillococcinum should be recommended to patients as an effective treatment for the flu.
   a. True
   b. False

8. The best recommendation for influenza protection is:
   a. American ginseng
   b. Asian ginseng
   c. Elderberry extract tea
   d. An annual influenza vaccine

9. Which species of ginseng may be useful for preventing the flu?
   a. American ginseng
   b. Asian ginseng
   c. Siberian ginseng
   d. a and b

10. What is the active ingredient in Sambucol?
    a. Oscillococcinum
    b. Asian ginseng
    c. Elderberry
    d. Siberian ginseng

Answers: 1-d; 2-a; 3-d; 4-c; 5-b; 6-b; 7-b; 8-d; 9-d; 10-c

REFERENCES

1. Jellin JM, Gregory PJ, Batz F, et al. Natural Medicines Comprehensive Database. Therapeutic Research Faculty. Stockton CA; 2004. Available at www.naturaldatabase.com. Accessed July 29, 2004.

2. Madav S, Tripathi HC, Tandan SK, et al. Antiallergic activity of andrographolide. *Indian J Pharm Sci.* 1998;60:176-8.

3. Puri A, Saxena R, Saxena RP, et al. Immunostimulant agents from *Andrographis paniculata*. *J Nat Prod.* 1993;56:995-9.

4. Caceres DD, Hancke JL, Burgos RA, et al. Use of visual analogue scale measurements (VAS) to assess the effectiveness of standardized *Andrographis paniculata* extract SHA-10 in reducing the symptoms of common cold. A randomized, double-blind, placebo study. *Phytomedicine.* 1999;6:217-23.

5. Melchior J, Palm S, Wikman G. Controlled clinical study of standardized *Andrographis paniculata* in common cold—a pilot trial. *Phytomedicine.* 1996;97;3:315-8.

6. Hancke J, Burgos R, Caceres D, et al. A double-blind study with a new monodrug Kan Jang: decrease of symptoms and improvement in the recovery from common colds. *Phytotherapy Res.* 1995;9:559-62.

7. Melchior J, Spasov AA, Ostrovskij OV, et al. Double-blind, placebo-controlled pilot and phase III study of activity of standardized *Andrographis paniculata* Herba Nees extract fixed combination (Kan Jang) in the treatment of uncomplicated upper-respiratory tract infection. *Phytomedicine.* 2000;7:341-50.

8. Gabrielian ES, Shukarian AK, Goukasova GI, et al. A double blind, placebo-controlled study of *Andrographis paniculata* fixed combination Kan Jang in the treatment of acute upper respiratory tract infections including sinusitis. *Phytomedicine.* 2002;9:589-97.

9. Poolsup N, Suthisisang C, Prathanturarug S, et al. *Andrographis paniculata* in the symptomatic treatment of uncomplicated upper respiratory tract infection: systematic review of randomized controlled trials. *J Clin Pharm Ther.* 2004;29:37-45.

10. Caceres DD, Hancke JL, Burgos RA, et al. Prevention of common colds with *Andrographis paniculata* dried extract: a pilot, double-blind trial. *Phytomedicine.* 1997;4:101-4.

11. Zhang CY, Tan BK. Mechanisms of cardiovascular activity of *Andrographis paniculata* in the anaesthetized rat. *J Ethnopharmacol.* 1997;56:97-101.

12. Zhao HY, Fang WY. Antithrombotic effects of *Andrographis paniculata* nees in preventing myocardial infarction. *Chin Med J* (Engl). 1991;104:770-5.

13. Amroyan E, Gabrielian E, Panossian A, et al. Inhibitory effect of andrographolide from *Andrographis paniculata* on PAF-induced platelet aggregation. *Phytomedicine.* 1999;6:27-31.

14. Barnes P, Powell-Griner E, McFann K, et al. CDC Advance Data Report #343. Complementary and Alternative Medicine Use Among Adults: United States, 2002.

15. Brinkeborn RM, Shah DV, Degenring FH. Echinaforce and other echinacea fresh plant preparations in the treatment of the common cold. A randomized, placebo controlled, double-blind clinical trial. *Phytomedicine.* 1999;6:1-6.

16. Lindenmuth GF, Lindenmuth EB. The efficacy of echinacea compound herbal tea preparation on the severity and duration of upper respiratory and flu symptoms: a randomized, double-blind, placebo-controlled study. *J Altern Complement Med.* 2000;6:327-34.

17. Dorn M, Knick E, Lewith G. Placebo-controlled, double-blind study of Echinaceae pallidae radix in upper respiratory tract infections. *Complement Ther Med.* 1997;5:40-2.

18. Henneicke-von Zepelin H, Hentschel C, Schnitker J, et al. Efficacy and safety of a fixed combination phytomedicine in the treatment of the common cold (acute viral respiratory tract infection): results of a randomised, double blind, placebo-controlled, multicentre study. *Curr Med Res Opin.* 1999;15:214-27.

19. Schulten B, Bulitta M, Ballering-Bruhl B, et al. Efficacy of *Echinacea purpurea* in patients with a common cold. A placebo-controlled, randomised, double-blind clinical trial. *Arzneimittelforschung.* 2001;51:563-8.

20. Goel V, Lovlin R, Barton R, et al. Efficacy of a standardized echinacea preparation (Echinilin) for the treatment of the common cold: a randomized, double-blind, placebo-controlled trial. *J Clin Pharm Ther.* 2004;29:75-83.

21. Grimm W, Muller HH. A randomized controlled trial of the effect of fluid extract of Echinacea purpurea on the incidence and severity of colds and respiratory infections. *Am J Med.* 1999;106:138-43.

22. Barrett BP, Brown RL, Locken K, et al. Treatment of the common cold with unrefined echinacea. A randomized, double-blind, placebo-controlled trial. *Ann Intern Med.* 2002;137:939-46.

23. Yale SH, Liu K. Echinacea purpurea therapy for the treatment of the common cold: a randomized, double-blind, placebo-controlled clinical trial. *Arch Intern Med.* 2004;164:1237-41.

24. Barrett B, Vohmann M, Calabrese C. Echinacea for upper respiratory infection. *J Fam Pract.* 1999;48:628-35.

25. Giles JT, Palat CT III, Chien SH, et al. Evaluation of echinacea for treatment of the common cold. *Pharmacotherapy.* 2000;20:690-7.

26. Melchart D, Linde K, Fischer P, et al. Echinacea for preventing and treating the common cold. *Cochrane Database Syst Rev.* 2000;2:CD000530.

27. Turner RB, Riker DK, Gangemi JD. Ineffectiveness of echinacea for prevention of experimental rhinovirus colds. *Antimicrob Agents Chemother.* 2000;44:1708-9.

28. Sperber SJ, Shah LP, Gilbert RD, et al. Echinacea purpurea for prevention of experimental rhinovirus colds. *Clin Infect Dis.* 2004;38:1367-71.

29. Mullins RJ, Heddle R. Adverse reactions associated with echinacea: the Australian experience. *Ann Allergy Asthma Immunol.* 2002;88:42-51.

30. Taylor JA, Weber W, Standish L, et al. Efficacy and safety of echinacea in treating upper respiratory tract infections in children: a randomized controlled trial. *JAMA.* 2003;290:2824-30.

31. Mullins RJ. Echinacea-associated anaphylaxis. *Med J Aust.* 1998;168:170-1.

32. Mullins RJ. Allergic reactions to echinacea. *J Allergy Clin Immunol.* 2000;104:S340-341;Abstract 1003.

33. Soon SL, Crawford RI. Recurrent erythema nodosum associated with echinacea herbal therapy. *J Am Acad Dermatol.* 2001;44:298-9.

34. Logan JL, Ahmed J. Critical hypokalemic renal tubular acidosis due to Sjogren's syndrome: association with the purported immune stimulant echinacea. *Clin Rheumatol.* 2003;22:158-9.

35. Lee AN, Werth VP. Activation of autoimmunity following use of immunostimulatory herbal supplements. *Arch Dermatol.* 2004;140:723-7.

36. Budzinski JW, Foster BC, Vandenhoek S, et al. An in vitro evaluation of human cytochrome P450 3A4 inhibition by selected commercial herbal extracts and tinctures. *Phytomedicine.* 2000;7:273-82.

37. Gorski JC, Huang S, Zaheer NA, et al. The effect of echinacea on CYP3A activity in vivo. *Clin Pharmacol Ther.* 2003;73(Abstract PDII-A-8):P94.

38. Martin NG, Carr AB, Oakeshott JG, et al. Co-twin control studies: vitamin C and the common cold. *Prog Clin Biol Res.* 1982;103:365-73.

39. Gorton HC, Jarvis K. The effectiveness of vitamin C in preventing and relieving the symptoms of virus-induced respiratory infections. *J Manipulative Physiol Ther.* 1999;22:530-3.

40. Carr AB, Einstein R, Lai LY, et al. Vitamin C and the common cold: using identical twins as controls. *Med J Aust.* 1981;2:411-2.

41. Audera C, Patulny RV, Sander BH, et al. Mega-dose vitamin C in treatment of the common cold: a randomized controlled trial. *Med J Aust.* 2001;175:359-62.

42. Hemila H. Vitamin C supplementation and common cold symptoms: factors affecting the magnitude of the benefit. *Med Hypotheses.* 1999;52:171-8.

43. Levine M, Rumsey SC, Daruwala R, et al. Criteria and recommendations for vitamin C intake. *JAMA.* 1999;281:1415-23.

44. Food and Nutrition Board, Institute of Medicine. Dietary Reference Intakes for Vitamin C, Vitamin E, Selenium, and Carotenoids. Washington, DC: National Academies Press, 2000. Available at www.nap.edu/books/0309069351/html.

45. Domingo JL, Gomez M, Llobet JM, et al. Effect of ascorbic acid on gastrointestinal aluminum absorption. *Lancet.* 1991;338:1467. Letter.

46. Partridge NA, Regnier FE, White JL, et al. Influence of dietary constituents on intestinal absorption of aluminum. *Kidney Int.* 1989;35:1413-7.

47. Slain D, Amsden JR, Khakoo RA, et al. Effect of high-dose vitamin C on the steady-state pharmacokinetics of the protease inhibitor indinavir in healthy volunteers. *Pharmacotherapy.* 2005;25:165-70.

48. Hume R, Johnstone JM, Weyers E. Interaction of ascorbic acid and warfarin. *JAMA.* 1972;219:1479.

49. Feetam CL, Leach RH, Meynell MJ. Lack of a clinically important interaction between warfarin and ascorbic acid. *Toxicol Appl Pharmacol.* 1975;31:544-7.

50. Weintraub M, Griner PF. Warfarin and ascorbic acid: lack of evidence for a drug interaction. *Toxicol Appl Pharmacol.* 1974;28:53-6.

51. Mossad SB, Macknin ML, Medendorp SV, et al. Zinc gluconate lozenges for treating the common cold. A randomized, double-blind, placebo-controlled study. *Ann Intern Med.* 1996;125:81-8.

52. Farr BM, Conner EM, Betts RF, et al. Two randomized controlled trials of zinc gluconate lozenge therapy of experimentally induced rhinovirus colds. *Antimicrob Agents Chemother.* 1987;31:1183-7.

53. Godfrey JC, Conant Sloane B, Smith DS, et al. Zinc gluconate and the common cold: a controlled clinical study. *J Int Med Res.* 1992;20:234-6.

54. Prasad AS, Fitzgerald JT, Bao B, et al. Duration of symptoms and plasma cytokine levels in patients with the common cold treated with zinc acetate. A randomized, double-blind, placebo-controlled trial. *Ann Intern Med.* 2000;133:245-52.

55. Petrus EJ, Lawson KA, Bucci LR, et al. Randomized, double-masked, placebo-controlled clinical study of the effectiveness of zinc acetate lozenges on common cold symptoms in allergy-tested subjects. *Curr Ther Res.* 1998;59:595-607.

56. Douglas RM, Miles HB, Moore BW, et al. Failure of effervescent zinc acetate lozenges to alter the course of upper respiratory tract infections in Australian adults. *Antimicrob Agents Chemother.* 1987;31:1263-5.

57. Weismann K, Jakobsen JP, Weismann JE, et al. Zinc gluconate lozenges for common cold. A double-blind clinical trial. *Dan Med Bull.* 1990;37:279-81.

58. Smith DS, Helzner EC, Nuttall CE Jr, et al. Failure of zinc gluconate in treatment of acute upper respiratory tract infections. *Antimicrob Agents Chemother.* 1989;33:646-8.

59. Jackson JL, Lesho E, Peterson C. Zinc and the common cold: a meta-analysis revisited. *J Nutr.* 2000;130:1512S-5S.

60. Marshall I. Zinc for the common cold. *Cochrane Database Syst Rev.* 2000;CD001364.

61. Mossad SB. Effect of zincum gluconicum nasal gel on the duration and symptom severity of the common cold in otherwise healthy adults. *QJM.* 2003;96:35-43.

62. Belongia EA, Berg R, Liu K. A randomized trial of zinc nasal spray for the treatment of upper respiratory illness in adults. *Am J Med.* 2001;111:103-8.

63. Hirt M, Nobel S, Barron E. Zinc nasal gel for the treatment of common cold symptoms: A double-blind, placebo-controlled trial. *Ear Nose Throat J.* 2000;79:778-82.

64. Turner RB. Ineffectiveness of intranasal zinc gluconate for prevention of experimental rhinovirus colds. *Clin Infect Dis.* 2001;33:1865-70.

65. Takkouche B, Regueira-Mendez C, Garcia-Closas R, et al. Intake of vitamin C and zinc and risk of common cold: a cohort study. *Epidemiology.* 2002;13:38-44.

66. Turner RB, Cetnarowski WE. Effect of treatment with zinc gluconate or zinc acetate on experimental and natural colds. *Clin Infect Dis.* 2000;31:1202-8.

67. Food and Nutrition Board, Institute of Medicine. Dietary Reference Intakes for Vitamin A, Vitamin K, Arsenic, Boron, Chromium, Copper, Iodine, Iron, Manganese, Molybdenum, Nickel, Silicon, Vanadium, and Zinc. Washington, DC: National Academies Press, 2000. Available at www.nap.edu/books/0309072794/html.

68. Barrett S. Zicam Marketers Sued. United States District Court Western District of Michigan Southern Division, Filed October 14, 2003, Case No. 4:03CV0146.

69. Jafek BW, Linschoten M, Murrow BW. Zicam Induced Anosmia. American Rhinologic Society 49th Annual Fall Scientific Meeting abstract. Orlando, Florida. September 20, 2003. Available at http://app.american-rhinologic.org/programs/2003ARSFallProgram071503.pdf. Accessed November 24, 2003.

70. DeCook CA, Hirsch AR. Anosmia due to inhalational zinc: a case report. *Chem Senses.* 2000;25:659. Abstract.

71. Barak V, Halperin T, Kalickman I. The effect of Sambucol, a black elderberry-based, natural product, on the production of human cytokines: I. Inflammatory cytokines. *Eur Cytokine Netw.* 2001;12:290-6.

72. Zakay-Rones Z, Varsano N, Zlotnik M, et al. Inhibition of several strains of influenza virus in vitro and reduction of symptoms by an elderberry extract (Sambucus nigra L.) during an outbreak of influenza B Panama. *J Altern Complement Med.* 1995;1:361-9.

73. Zakay-Rones Z, Thom E, Wollan T, et al. Randomized study of the efficacy and safety of oral elderberry extract in the treatment of influenza A and B virus infections. *J Int Med Res.* 2004;32:132-40.

74. Luo P, Wang L. Peripheral blood mononuclear cell production of TNF-alpha in response to North American ginseng stimulation. *Alt Ther.* 2001;7:S21. Abstract.

75. Wang M, Guilbert LJ, Ling L, et al. Immunomodulating activity of CVT-E002, a proprietary extract from North American ginseng (Panax quinquefolium). *J Pharm Pharmacol.* 2001;53:1515-23.

76. Predy GN, Goel V, Lovlin R, et al. Efficacy of an extract of North American ginseng containing poly-furanosyl-pyranosyl-saccharides for preventing upper respiratory tract infections: a randomized controlled trial. *CMAJ.* 2005;173:1043-8.

77. Scaglione F, Cattaneo G, Alessandria M, et al. Efficacy and safety of the standardized ginseng extract G115 for potentiating vaccination against the influenza syndrome and protection against the common cold. *Drugs Exp Clin Res.* 1996;22(2):65-72.

78. McElhaney JE, Gravenstein S, Cole SK, et al. A placebo-controlled trial of a proprietary extract of North American ginseng (CVT-E002) to prevent acute respiratory illness in institutionalized older adults. *J Am Geriatr Soc.* 2004;52:13-9.

79. Caron MF, Hotsko AL, Robertson S, et al. Electrocardiographic and hemodynamic effects of Panax ginseng. *Ann Pharmacother.* 2002;36:758-63.

80. Vuksan V, Stavro MP, Sievenpiper JL, et al. Similar postprandial glycemic reductions with escalation of dose and administration time of American ginseng in type 2 diabetes. *Diabetes Care.* 2000;23:1221-6.

81. Sotaniemi EA, Haapakoski E, Rautio A. Ginseng therapy in non-insulin dependent diabetic patients. *Diabetes Care.* 1995;18:1373-5.

82. Janetzky K, Morreale AP. Probable interaction between warfarin and ginseng. *Am J Health Syst Pharm.* 1997;54:692-3.

83. Yuan CS, Wei G, Dey L, et al. American ginseng reduces warfarin's effect in healthy patients: a randomized, controlled trial. *Ann Intern Med.* 2004;141:23-7.

84. Vickers AJ, Smith C. Homoeopathic Oscillococcinum for preventing and treating influenza and influenza-like syndromes. *Cochrane Database Syst Rev.* 2004;1:CD001957.

CHAPTER 14

Weight Loss

Karen Shapiro

Weight-loss products appear to have gotten more confusing than cough and cold products. The number of choices with fantastic claims is bewildering. Marketing is designed to appeal to a result-oriented, pill-popping culture. Consumers want fast results and are willing to believe in and spend lots of money on these products.

Much confusion is removed by breaking products into categories of agents, similar to the way cough and cold products are broken down. Despite fancy packaging to the contrary, many of the basic ingredients are the same. In addition to common ingredients, products sometimes contain other agents in an attempt to gain a marketing edge. The use of these agents may be based on questionable study data or be based on nothing at all.

KEY POINTS

- Caffeine is present in most combination weight-loss products.
- Common sources of caffeine include green and black tea, cola nut, guarana, yerba mate, and cocoa.
- High doses of caffeine aggravate hypertension and anxiety.
- Ceasing caffeine consumption causes withdrawal symptoms.
- Ephedra-like substances are present in most combination weight-loss products.
- Bitter orange, the most common ephedra-like substance, contains synephrine.
- Ephedra-like substances should be avoided; patients develop rapid tolerance to these agents which can become dangerous in excessive doses.
- Caffeine and ephedra-like substances are often used together for additive effects.
- Calcium intake contributes to a healthy weight.
- Potassium may be present for a mild diuretic effect. This may be dangerous in patients with impaired renal function and in those taking agents that increase potassium retention.
- Chromium, alpha-lipoic acid, glucomannan, and vanadium are common ingredients, all of which do not contribute to weight loss.
- A proprietary blend of white kidney bean extract is used in most starch-blocking supplements.
- The white kidney bean extract may contribute to modest weight loss; the doses used in these products are larger and have not been studied.

Many of the agents discussed here have been studied for use as glucose-lowering agents in patients with diabetes—a condition in which many patients are overweight. A modest reduction in blood glucose, if this does occur, does not contribute to weight loss. Other agents present in the combination products are used for cholesterol reduction. Lowering cholesterol, which will not occur anyway with the typical doses used in these products, does not produce weight loss.

Sometimes, a product may produce short-term weight loss but does not provide long-term benefit, or the long-term health consequences may be harmful. The primary mechanism of action for almost all weight-loss products is the inclusion of ephedra-like stimulants and excessive doses of caffeine. A person can lose weight short term by taking pseudoephedrine (Sudafed®) cold tablets, with or without caffeine, but it is certainly not healthy and the weight will quickly reappear.

Another concern involves products that lump individual agents into "proprietary" blends, where quantity is given for the blend only. This makes

PRODUCT	INGREDIENTS
Hydroxycut® Advanced Weight Loss Formula—Ephedra Free	Calcium 265 mg, chromium 133 mcg, potassium 249 mg, and 1.7 g Hydroxagen plus proprietary blend, includes garcinia camboga, L-carnitine, alpha lipoic-acid, glucomannan, willow bark extract, **green leaf tea extract, caffeine, and guarana**
Dexatrim® Natural Ephedrine Free Formula	Chromium 83 mcg and 340 mg of proprietary blend, includes vanadium, **green leaf tea extract, cola nut extract with added caffeine, guarana seed extract with added caffeine,** *bitter orange 120 mg,* Siberian ginseng, fenugreek seed extract, ginger root, and licorice root
Xenadrine® EFX Ephedrine-Free Advanced Thermogenic/Energy Formula	Vitamin C 100 mg, vitamin B6 10 mg, pantothenic acid 12 mg, magnesium 10 mg, and 1415 mg of Thermodyne proprietary complex, includes **cocoa extract, yerba mate,** *bitter orange,* L-tyrosine, acetyl-L-tyrosine, pentahydro- and tetrahydrooxyflavone, DMEA (2-dimethylaminoethanol), grape seed extract
Carb Intercept®	White kidney bean extract 1 g

Bold = contains caffeine   ***Bold italic*** = contains synephrine

dose–response relationships for the individual agents impossible to determine and goes against the grain of an evidence-based practitioner.

But that is the nature of the beast—weight-loss agents are not evidence-based, where recommendations are based on the quality of the evidence. Dr. Phil McGraw, well-known talk show host and self-appointed weight-loss expert, has a line of weight-loss supplements based on body shape called Dr. Phil's Shape Up!® weight management supplement and multivitamin for apple or pear body types. He was questioned about the applicability of his apple-versus-pear theory, the "logic" behind his supplement marketing. Dr. Phil stated that his medical and nutrition advisers "felt it was an important distinction to make." In the end, he said, he did not know whether the apple-versus-pear theory would hold up. "I think it's important to note that it may or may not," he acknowledged.[1]

Practitioners and patients can enter this mystical world with self-confidence in their ability to see through product claims—even better than Dr. Phil. The laws of thermodynamics have not changed which means that energy cannot be created or destroyed but can only be transformed from one form into another. When extra calories are eaten and energy expenditure is not increased, that energy is going to be stored somewhere, most likely in the "apple" middle and/or "pear" rear.

If a miracle existed, it would be known. With more concern about overweight and obesity than ever and more products and services marketed to treat these problems, 66% of Americans are overweight, up from 43% in the early 1940s. Thirty-percent of American adults are obese (BMI over 30).[2] There is no

PATIENT CASE

AMY

Amy is a 45-year-old Caucasian female. She is in good health. Her blood pressure is normal. Her cholesterol panel is within normal limits. Her height is 5'2", weight 122 lbs. She has two children, ages 11 and 9. Amy reports that she used to be thin and is upset that the weight has crept on over the years. She wants to get back to her college weight of 110 pounds.

Amy is a professional who works in an office and travels 20% of the time. She describes her job as "high stress." She eats breakfast in her car and grabs a quick lunch from whatever restaurant she is near, except if she has to eat out for work. Dinner is whatever she or her husband pick up on the way home that the kids will eat. She exercises at a gym one or two times a week. She does not smoke and consumes red wine socially. She is considering buying a weight-loss product and asks for advice.

scientific evidence that supplements help anyone lose weight—except with the use of stimulants where the weight loss is dangerous and ephemeral.

Combination products are discussed in this chapter, along with one product marketed as an individual agent. Weight loss agents are usually marketed as combination products. It is necessary to be able to recognize and understand the use of the common ingredients used in the combination products.

Caffeine and the ephedra-like stimulants are discussed initially. These represent the primary active ingredients in the first three products. Additional agents in the combination products are discussed next. The last product presented here is an extract from white kidney beans present in the carb-blocking products. See the accompanying table.

CAFFEINE-CONTAINING PRODUCTS

Natural products with significant caffeine content include:[3]

- Green and black tea
- Cola nut (used in cola drinks)
- Guarana (the substitute for cola drinks in Brazil)
- Yerba mate (also called mate, this product contains caffeine, theophylline, and theobromine)
- Cocoa (used to make chocolate)

Caffeine acts as a CNS stimulant with a dose-dependent response. How much one consumes determines the effect on the body. Low or normal doses in healthy individuals pose little health risk. At low doses, caffeine helps a person wake up and produces more rapid and clear thought. At higher doses, caffeine can produce increases in heart rate and force of contraction. This increases the workload on the heart and can compromise patients with decreased cardiac function or arrhythmias. High doses aggravate hypertension and anxiety. Other effects from caffeine are elevated levels of gastric acid, pepsin, blood glucose, renin, and free fatty acids. Ceasing the consumption of caffeine causes withdrawal symptoms, such as headache and fatigue.

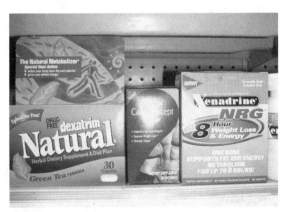

Weight-loss products

How much caffeine, in milligrams (mg), is present in these products? According to the manufacturers, each serving of the Dexatrim formula contains 80 mg of caffeine[4], a Xenadrine serving is equivalent to one cup of coffee,[5] and a Hydroxycut serving is equivalent to two cups of coffee.[6] Converting cups of coffee to milligrams of caffeine is tricky. A typical cup of brewed coffee has about 135 mg caffeine—but a cup of Starbuck's coffee contains more caffeine than a cup of coffee from the International House of Pancakes. Amount of caffeine, when not specifically specified, could be anywhere from 30 to 200 mg per cup.[3]

One source of listed caffeine is green tea, which may be beneficial in maintaining a healthy weight.[7,8] Green tea may increase thermogenesis (raising the body's heat production) via beta-adrenoceptor activation. The beta-adrenergic receptors are sympathetic nervous system receptors that are activated by norepinephrine and cause the reactions present in the fight or flight response, such as blood vessel constriction and increased heart rate and temperature. The increase in body energy expenditure could be very slight and have no noticeable outward effect (at least in the present), yet contribute over time to weight loss.

Green and black tea contain compounds called flavonoids that have important health benefits, similar to the compounds found in red wine. Flavonoids are powerful antioxidants and have immune-boosting properties.[9] They can help protect against cardiovascular disease and may protect against certain types of cancer.[10-12] Drinking tea can be a wise addition to a healthy lifestyle. Consuming excessive amounts of caffeine in the form of combination products that contain unhealthy and/or untested combinations of products is not a wise decision.

EPHEDRA-LIKE STIMULANTS

The FDA moved to ban ephedra in weight loss products in April 2004 after there were 16,000 reported adverse events and numerous deaths. Ephedra and its counterparts stimulate the central nervous system, increasing blood pressure and heart rate. Chronic use can lead to tolerance and dependence, requiring ever-larger doses to obtain the same effects and increasing the risk of serious consequences. Ephedra was banned, but the use of over-the-counter products containing ephedrine or other stimulants similar to ephedra were not included in the ban.

Bitter orange is the most popular ephedra-like agent in current use. The peel from the bitter orange contains a chemical called synephrine, a substance similar to ephedra and pseudoephedrine. Bitter orange may be banned in the future. In the meantime, patients should be advised not to use it. The

Dexatrim and Xenadrine products, both advertised as ephedrine-free, contain synephrine and should not be considered to be safer alternatives. These products also have high caffeine content. Some of the "ephedrine-free" products do not contain synephrine and instead have multiple sources of caffeine (e.g., the Hydroxycut product.)

HYDROXYCUT ADDITIONAL AGENTS

Garcinia is an extract from the *Garcinia cambogia* tree. The active ingredient in garcinia is hydroxycitric acid (HCA). This is the main ingredient in the "diet pills" CitriMax, MAGTAB Hydroxycitric Acid, and others. The use of hydroxycitric acid is based upon weight loss in animal studies. Hydroxycitric acid blocks an enzyme necessary for fatty acid synthesis.[13] The mechanism is accurate; however, this does not produce weight loss.

Calories do not disappear because they cannot be converted to fat; they are directed into other metabolic pathways, such as storage in the form of glycogen. Second, this mechanism assumes that the problem with weight gain has to do with fat synthesis. It is not the case: humans do not synthesize too much fat but rather consume too much. The preliminary data obtained from animal studies have never been extended to humans. Studies using human subjects found no significant difference in the amount of weight lost between the groups receiving garcinia and those getting a placebo.[14,15] This product represents a good example of twisting available data for a marketing advantage. Hydroxycitric acid appears as safe as citric acid.[16] There are no known drug interactions.

L-Carnitine is an amino acid synthesized by the body from the amino acids lysine and methionine and is important in fatty acid metabolism. There is no common dietary insufficiency and supplementation is required only if an inherited metabolic disorder exists or in a few specific diseases that cause low serum levels, including end-stage renal disease. Supplemental carnitine may be useful in certain types of cardiac disease and for toxicities caused by some prescription drugs.[17-19]

L-Carnitine is present in many products used for performance enhancement and muscle building. Yet, L-carnitine is not useful for exercise performance or weight loss.[20,21] Once L-carnitine became prevalent in muscle-building formulas, marketers of weight-loss products began to use this angle for promoting their products. A book called *The Carnitine Miracle* helped fuel a carnitine craze. It claims, among other things, that L-carnitine is "the most important nutrient for naturally supporting the weight loss process."[22] In general, the side effects of L-carnitine are few. Nausea, diarrhea, and other gastrointestinal problems occur. This product should be avoided in patients using warfarin, due to possible potentiation of the anticoagulant effect.[23]

Calcium is another important agent for good health that is often present in weight-loss products. Like tea, the benefits from calcium were not found in the marketed products discussed here but from taking calcium itself. A diet rich in calcium may aid weight loss.[24] Research has looked at the long-term weight of women with higher calcium intake. It appears that women who consume more calcium are less likely to be obese.[25] In another study, increasing dietary calcium significantly augmented weight and fat loss. Dairy products had a greater effect than taking calcium in the form of supplements.[26]

The mechanism for calcium's role in weight loss is not clear but it may be possible that fat cells burn more rapidly when there is a higher intracellular calcium content.[27] If it is verified that dairy sources have a greater effect on weight loss than calcium supplied by supplements, then it can be postulated that there are compounds in dairy which are contributing to the weight loss. If dairy sources are used, the calcium should be present in low-fat sources and total calorie intake must be considered. If calcium food products are added to a diet without reducing calories somewhere else (i.e., the total caloric intake increases), the person will put on weight.

Calcium, when taken in normal doses, is well tolerated. Some patients experience stomach upset. Clinicians often tell patients that calcium causes constipation—and while this may occur in individual patients, it has not been shown to occur in population studies.[28] Many patients are chronically constipated, but this is most often due to poor fluid intake and a low-fiber diet.

Chromium plays an essential role in the metabolism of carbohydrates and lipids and in the action of insulin. Chromium, when taken in certain salt forms and in relatively large doses, helps reduce insulin resistance and may reduce lipids in persons with diabetes, but there is no evidence that chromium contributes to weight loss.[29]

Potassium is the next ingredient in the Hydroxycut product. Potassium is the major intracellular cation and is essential for proper functioning of all cells and nervous tissue. For good health, the diet should be rich in potassium. Consuming potassium-rich foods is part of the Dietary Approaches to Stop Hypertension (DASH) eating plan.[30] However, **supplemental potassium should not be consumed** unless a physician has taken a serum level and determined that this is required. Too much potassium is dangerous—high potassium levels induce arrhythmias and can result in cardiac arrest. Added potassium can be dangerous in patients with decreased kidney function or in those using multiple agents that cause potassium retention.

How much potassium is in this supplement? This is not known for sure because there are two salts present and the atomic weight of each and proportion of each would be needed to calculate the amount of milliequivalents of potassium. An approximate guess is 4 mEq per serving or 12 mEq daily if the package instructions are followed.

The potassium is likely present for a diuretic effect, similar to the mechanism of action of the potassium-retaining prescription drug triamterene. This drug exerts a diuretic effect on the distal renal tubules of the kidneys by inhibiting sodium reabsorption in exchange for potassium and hydrogen ions. If this occurs, it is not weight loss. It is water loss. And it is not safe to lose weight by depleting fluid volume. Weight loss in this manner is also not sustainable.

Alpha-lipoic acid, a potent antioxidant, is present in this and many other weight-loss products. Studies have shown that this agent may help with glucose control in patients with diabetes.[31,32] There is no evidence that alpha-lipoic-acid is useful for weight loss.

Glucomannan is a high-fiber product with a mechanism of action for cholesterol reduction and minor improvements in glucose control similar to that of oats and psyllium (see Chapter 10).[33] The quantity of glucomannan fibers (which are bulky and are usually taken as several grams daily) would be negligible in any combination weight loss product.

Willow bark extract contains salicylates and is typically used as an analgesic, similar to aspirin. Persons who are allergic to aspirin should not use this product. Willow bark can cause gastrointestinal distress and bleeding, however the amount present is negligible.

DEXATRIM NATURAL ADDITIONAL AGENTS

Vanadium, an element and a trace mineral, is another agent that may be useful in improving insulin sensitivity.[34] There is no evidence that vanadium is useful for weight loss. At lower doses vanadium is well tolerated. **Siberian ginseng** has been studied for use in diabetes but data for American ginseng as a hypoglycemic agent are stronger.[35,36] Siberian ginseng is likely present because it is cheaper than other ginseng formulations. Side effects are uncommon and not expected with the amount included in this product. **Fenugreek** seed is another component with a limited amount of data for diabetes and cholesterol reduction and none for weight loss.[37,38] At the low doses that could be contained in this supplement, side effects should not be a problem. At higher doses, fenugreek is used as a bulk fiber and causes diarrhea and flatulence and can adsorb certain oral medications. **Ginger** is used for nausea (primarily) and a few other conditions but not for weight loss. It is safe in typical doses. **Licorice** is the last component of Dexatrim Natural. There are no data that licorice is useful for any health condition. Excessive use of licorice causes several health problems, including hypokalemia, elevated blood pressure, and inhibition of hepatic enzymes, but these would not be expected to occur with the amount of licorice likely present in this supplement.[39]

Xenadrine Additional Agents

The **B vitamins** included are in healthy amounts but are not known to be useful for weight loss. L-Tyrosine is a nonessential amino acid that has no known data for weight loss. The **"oxyflavone"** compounds likely refer to quercetin, a flavonoid found in many food products, including some fruits and red wine. This type of flavonoid is considered healthy but is not known to contribute to weight loss. **DMEA** stands for 2-dimethylaminoethanol, a precursor to choline. This product is not known to help with weight loss. It has been studied for use as an adjunctive agent for Alzheimer's disease as a "booster" of acetylcholine; however, this use was not shown to be beneficial.[40] **Grape seed extract** is an antioxidant which is useful for general health. There is no known benefit for using grape seed extract for weight loss.

Carb Intercept

This product is simple to discuss because it contains just one ingredient, a white kidney bean extract that works as a "starch blocker" by inhibiting the digestive enzyme alpha-amylase. Amylases break down starches, including potatoes, wheat, and corn, into smaller sugars, which can then be absorbed and eventually converted into individual glucose units. Blocking this enzyme could reduce or delay absorption of consumed starches. Cruder versions of white kidney bean extracts were first marketed in the 1970s. These were found to be ineffective and disappeared from store shelves.[41]

The agent in the products marketed now, including Carb Intercept, Carb Blocker®, and others, contain an extract that is more refined and concentrated. This extract is marketed as a proprietary formula called Phaseolamin 2250, or Phase 2. A well-designed but small 8-week study showed an average weight loss of 3.79 pounds in the Phase 2 group versus a 1.65 pound average weight loss in the placebo group. The difference was not statistically significant. Triglyceride levels were significantly improved in the Phase 2 group. In this study, the extract was dosed at 1500 mg twice daily and was well tolerated. Carb Intercept is dosed much higher—at 2 g with each carbohydrate-containing meal. Gastrointestinal side effects would likely be present at higher doses. Undigested starch is not absorbed and would be expected to produce flatulence and diarrhea.[42] The makers of some of the starch-blocking supplements were forced by the FDA to modify their claims that you can eat what you want and still lose weight.[43]

Chitosan, another so-called "fat blocker" that is present in other products, purportedly works by binding fats in the intestine. This product has no proven efficacy for weight loss.[44] Chitosan is made from shellfish and could possibly cause a serious reaction in people with shellfish allergies.

PATIENT DISCUSSION

Is Amy overweight? That depends who is asked. Not according to health care professionals, except for those in the cosmetic surgery business. The primary issue of overweight and obesity concerns health risk. Her BMI is 22.3, within the normal range of 18.5 to 24.9. She would be overweight if her BMI was greater than 25, and obese at a BMI of 30 or more. When she weighed 110 pounds, her BMI was 20.1, still a healthy value and more acceptable to the patient.

Amy represents the typical patient in that she does not have sustained physical activity in her daily routine and is likely consuming more calories than she expends on a regular basis.[45,46] Losing weight is not the subject of this chapter and there are free, reputable resources available for patients who need assistance with weight management.[47]

The bottom line for Amy if she wishes to maintain healthy, long-term weight loss is to change the way she is living day-to-day. A helpful suggestion would be to steer her out of the supplement aisle and into the gym more frequently. Five-hundred calories a day in physical activity is 1 pound weight loss weekly (500 kcal (7 days = 3,500 kcal, or the equivalent of 1 pound.) This is in addition to the other health benefits from regular physical activity and may be all Amy needs to return to her earlier weight.

If Amy wants to speed the process along, or for patients who need to lose more weight (and regularly over-consume), the diet can be reduced by 500 kcal (or more, if applicable) daily. This would increase Amy's weight loss to approximately 2 pounds weekly. If she loses weight faster than this, whether via the use of excessive caffeine, other stimulants, or drastic calorie reduction, the weight loss will not be sustainable. If she wants to take something, the clinician could recommend adequate calcium consumption and consuming tea as healthy additions to her diet.

SELF-ASSESSMENT

1. Which of the following are common sources of caffeine in weight loss products?
   a. Guarana
   b. Cola nut
   c. Green or black tea
   d. All of the above

2. Which of the following contains synephrine, a product similar to ephedra?
   a. Yerba mate
   b. L-Tyrosine
   c. Bitter orange
   d. DMEA

3. Which of the following products contains healthy flavonoids and may contribute to an increase in thermogenesis?
   a. Green tea
   b. Chromium
   c. L-Carnitine
   d. Fenugreek

4. Which is TRUE concerning calcium for weight loss:
   a. Calcium supplements offer more benefit than calcium consumed in dairy products.
   b. The intracellular calcium concentration may affect weight loss.
   c. Calcium should be consumed at doses greater than 2 g/day.
   d. Most Americans get too much calcium and supplements should be avoided.

5. Problems with ingredients in "proprietary blends" used in weight loss products include:
   a. Product content is difficult to determine.
   b. Individual agents may not be efficacious for weight-loss.
   c. Study data are often questionable.
   d. All of the above.

6. Which is TRUE concerning excessive use of caffeine:
   a. Symptoms of anxiety may worsen.
   b. Heart rate can increase.
   c. Withdrawal symptoms will be present when the product is discontinued.
   d. All of the above.

7. Which of the following agents contains salicylates?
   a. DMEA
   b. L-Tyrosine
   c. Willow bark
   d. L-Carnitine

8. A 44-year-old white male presents with complaints of excess weight. He has no known conditions except for gout. He has a BMI of 26. He is sedentary and consumes mostly fast-food, including burgers and soft drinks. He states he has been overweight most of his adult life. The best option for this patient is to recommend:
   a. A healthy diet plan with controlled caloric intake used in conjunction with increased physical activity
   b. One of the Dexatrim products
   c. Carb Intercept by Natrol
   d. Xenadrine EFX Ephedrine-Free Formula

9. CitriMax, Hydroxycut, and other products contain hydroxycitric acid (HCA). This product is known to contribute modestly to weight loss.
   a. True
   b. False

10. Carb Intercept, Carb Blocker, and other products contain an ingredient extracted from white kidney beans. This agent works by the following mechanism:
    a. It inhibits the digestive enzyme alpha-amylase.
    b. It contains caffeine and increases thermogenesis.
    c. It contains synephrine and causes a "stimulant" like effect.
    d. It decreases appetite.

Answers: 1-d; 2-c; 3-a; 4-b; 5-d; 6-d; 7-c; 8-a; 9-b; 10-a

REFERENCES

1. Burros M, Day S. Doubt Cast on Food Supplements for Weight Control. *NY Times.* Oct. 27, 2003.

2. Hedley AA, Ogden CL, Johnson CL, et al. Prevalence of overweight and obesity among US children, adolescents, and adults, 1999-2002, *JAMA.* 2004; 291:2847-50.

3. Durant KL. Known and hidden sources of caffeine in drug, food and natural products. *J Am Pharm Assoc.* 2002;42:625-37.

4. Available at www.dexatrim.com/product_noeph.asp. Accessed July 26, 2004.

5. Available at www.cytodyne.com/products.html?product_id=56&site= xenadrine. Accessed July 26, 2004.

6. Available at www.hydroxycut.com/WOMEN/FAQS/NEW_HYD/new_ hydroxycut_faq_02.shtml. Accessed July 26, 2004.

7. Chantre P, Lairon D. Recent findings of green tea extract AR25 (Exolise) and its activity for the treatment of obesity. *Phytomedicine.* 2002;9:3-8.

8. Choo JJ. Green tea reduces body fat accretion caused by high-fat diet in rats through beta-adrenoceptor activation of thermogenesis in brown adipose tissue. *J Nutr Biochem.* 2003 Nov;14(11):671-6.

9. Ahmad N, Katiyar SK, Mukhtar H. Antioxidants in chemoprevention of skin cancer. *Curr Probl Dermatol.* 2001;29:128-39.

10. Cao Y, Cao R, Brakenhielm E. Antiangiogenic mechanisms of diet-derived polyphenols. *J Nutr Biochem.* 2002 Jul;13(7):380-90.

11. Duffy SJ, Keaney JF Jr, Holbrook M, et al. Short- and long-term black tea consumption reverses endothelial dysfunction in patients with coronary artery disease. *Circulation.* 2001 Jul 10;104(2):151-6.

12. Geleijnse JM, Launer LJ, Van der Kuip, et al. Inverse association of tea and flavonoid intakes with incident myocardial infarction: the Rotterdam Study. *Am J Clin Nutr.* 2002 May;75(5):880-6.

13. Jena BS, Jayaprakasha GK, Singh RP, et al. Chemistry and biochemistry of (-)-hydroxycitric acid from Garcinia. *J Agric Food Chem.* 2002 Jan 2;50(1):10-22. Review.

14. Kovacs EM, Westerterp-Plantenga MS, Saris WH. The effects of 2-week ingestion of (-)-hydroxycitrate and (-)-hydroxycitrate combined with medium-chain triglycerides on satiety, fat oxidation, energy expenditure and body weight. *Int J Obes Relat Metab Disord.* 2001;25:1087-94.

15. Heymsfield SB, Allison DB, Vasselli JR, et al. Garcinia cambogia (hydroxycitric acid) as a potential antiobesity agent: a randomized controlled trial. *JAMA.* 1998 Nov 11;280(18):1596-600.

16. Soni MG, Burdock GA, Preuss HG, et al. Safety assessment of (-)-hydroxycitric acid and Super CitriMax®, a novel calcium/potassium salt. *Food Chem Toxicol.* 2004 Sep;42(9):1513-29.

17. Ghidini O, Azzurro M, Vita G, et al. Evaluation of the therapeutic efficacy of L-carnitine in congestive heart failure. *Int J Clin Pharmacol Ther Toxicol.* 1988;26:217-20.

18. Cherchi A, Lai C, Angelino F, et al. Effects of L-carnitine on exercise tolerance in chronic stable angina: a multicenter, double-blind, randomized, placebo-controlled, crossover study. *Int J Clin Pharmacol Ther Toxicol.* 1985;23:569-72.

19. Murakami K, Sugimoto T, Woo M, et al. Effect of L-carnitine supplementation on acute valproate intoxication. *Epilepsia.* 1996;37:687-9.

20. Marconi C, Sassi G, Carpinelli A, et al. Effects of L-carnitine loading on the aerobic and anaerobic performance of endurance athletes. *Eur J Appl Physiol Occup Physiol.* 1985;54(2):131-5.

21. Colombani P, Wenk C, Kunz I, et al. Effects of L-carnitine supplementation on physical performance and energy metabolism of endurance-trained athletes: a double-blind crossover field study. *Eur J Appl Physiol Occup Physiol.* 1996;73(5):434-9.

22. Crayton R. Carnitine: the powerhouse nutrient. *Better Nutrition.* May 1999, p. 34.

23. Bachmann HU, Hoffmann A. Interaction of food supplement L-carnitine with oral anticoagulant acenocoumarol. *Swiss Med Wkly.* 2004;134:385.

24. Teegarden D. Calcium intake and reduction in weight or fat mass. *J Nutr.* 2003;133:249S-51S.

25. Zemel MB, Richards J, Milstead A, et al. Effects of calcium and dairy on body composition and weight loss in African-American adults. *Obes Res.* 2005 Jul;13(7):1218-25.

26. Calcium and dairy acceleration of weight and fat loss during energy restriction in obese adults. *Obes Res.* 2004 Apr;12(4):582-90.

27. Zemel MB, Thompson W, Milstead A, et al. Calcium and dairy acceleration of weight and fat loss during energy restriction in obese adults. *Obes Res.* 2004 Apr;12(4):582-90.

28. Clemens JD, Feinstein AR. Calcium carbonate and constipation: a historical review of medical mythopoeia. *Gastroenterology.* 1977;72:957-61.

29. Trent LK, Thieding-Cancel D. Effects of chromium picolinate on body composition. *J Sports Med Phys Fitness.* 1995;35:273-80.

30. Available at www.nhlbi.nih.gov/health/public/heart/hbp/dash/. Accessed July 30, 2004.

31. Jacob S, Henriksen EJ, Schiemann AL, et al. Enhancement of glucose disposal in patients with type 2 diabetes by alpha-lipoic acid. *Arzneimittelforschung.* 1995;45:872-4.

32. Jacob S, Ruus P, Hermann R, et al. Oral administration of RAC-alpha-lipoic acid modulates insulin sensitivity in patients with type-2 diabetes mellitus: a placebo-controlled, pilot trial. *Free Rad Biol Med.* 1999;27:309-14.

33. Walsh DE, Yaghoubian V, Behforooz A. Effect of glucomannan on obese patients: a clinical study. *Int J Obes.* 1984;8:289-93.

34. Cohen N, Halberstam M, Shlimovich P, et al. Oral vanadyl sulfate improves hepatic and peripheral insulin sensitivity in patients with non-insulin-dependent diabetes mellitus. *J Clin Invest.* 1995;95:2501-9.

35. Hikino H, Takahashi M, Otake K, et al. Isolation and hypoglycemic activity of eleutherans A, B, C, D, E, F, and G: glycans of Eleutherococcus senticosus roots. *J Nat Prod.* 1986;49:293-7.

36. Vuksan V, Stavro MP, Sievenpiper JL, et al. Similar postprandial glycemic reductions with escalation of dose and administration time of American ginseng in type 2 diabetes. *Diabetes Care*. 2000;23:1221-6.

37. Madar Z, Abel R, Samish S, et al. Glucose-lowering effect of fenugreek in non-insulin dependent diabetics. *Eur J Clin Nutr*. 1988;42:51-4.

38. Bhardwaj PK, Dasgupta DJ, Prashar BS, et al. Control of hyperglycaemia and hyperlipidaemia by plant product. *J Assoc Physicians India*. 1994;42:33-5.

39. Zhou S, Koh HL, Gao Y, et al. Herbal bioactivation: the good, the bad and the ugly. *Life Sci*. 2004 Jan 9;74(8):935-68.

40. Ferris SH, Sathananthan G, Gershon S, et al. Senile dementia: treatment with deanol. *J Am Geriatr Soc*. 1977;25:241-4.

41. Carlson GL, Li BU, Bass P, et al. A bean alpha-amylase inhibitor formulation (starch blocker) is ineffective in man. *Science*. 1983;219:393-5.

42. Boivin M, Zinsmeister AR, Go VL, et al. Effect of a purified amylase inhibitor on carbohydrate metabolism after a mixed meal in healthy humans. *Mayo Clin Proc*. 1987;62:249-55.

43. Warning Letter for Weight Loss Products. Available at http://vm.cfsan.fda.gov/~dms/wl-ltr6.html. Accessed July 27, 2004.

44. Ho SC, Tai ES, Eng PH, et al. In the absence of dietary surveillance, chitosan does not reduce plasma lipids or obesity in hypercholesterolaemic obese Asian subjects. *Singapore Med J*. 2001;42:6-10.

45. Nielsen SJ, Siega-Riz AM, Popkin BM. Trends in energy intake in U.S. between 1977 and 1996: similar shifts seen across age groups. *Obes Res*. 2002;10:370-8.

46. Nielsen SJ, Popkin BM. Patterns and trends in food portion sizes, 1977-1998. *JAMA*. 2003;289:450-3.

47. National Heart, Lung, and Blood Institute. Clinical guidelines on the identification, evaluation, and treatment of overweight and obesity in adults. 1998. Available at www.nhlbi.nih.gov/guidelines/index.htm. Accessed July 12, 2005.

CHAPTER 15
Performance Enhancement

*Julia Ireland, Sandra Shibuyama,
Leila Khajehmolaei, and James D. Scott*

Training for athletic event performance and body-building entail increasing lean body mass (muscle and bone) and improving endurance. Protein is the main body-building material for increasing muscle mass, and protein supplementation, up to a point, can be useful for both body-building and athletic performance. Energy can be obtained from any of the three main components of food: fats, proteins, and carbohydrates.

Protein and carbohydrates are the preferred energy sources. And, as will be discussed, both of these can potentially improve athletic performance, speed

KEY POINTS

- Protein and carbohydrates are preferred dietary energy sources.
- Protein intake should be primarily from dietary sources, such as animal products and combinations of vegetarian foods.
- Protein supplements include soy and whey.
- Branched-chain amino acid supplementation can be beneficial to critically ill and dialysis patients.
- Evidence does not support the use of branched-chain amino acids, glutamine, or L-carnitine for improving athletic performance.
- Glutamine is used as a nutritional supplement in patients with trauma and HIV/AIDS wasting syndrome.
- L-Carnitine is involved in the conversion of fats for energy and in many other cellular processes.
- Creatine supplements improve muscle performance in short, intense workouts.
- Creatine supplements cause fluid retention.
- Adequate carbohydrate and fluid intake during training is essential.
- Sport drinks provide needed energy and fluids.
- Carbohydrates consumed before, during, and after strenuous exercise can help prevent breakdown of muscle protein.
- Antioxidant supplements, in safe doses, can be recommended.
- Individual athletes, dependent on dietary intake and other factors, may require calcium and magnesium supplementation.
- Iron, in the low doses present in many multivitamins, can be used to help prevent the anemia that can develop in serious athletes.

Product	Dosage	Effect	Safety Concerns
Protein supplements containing soy or whey	0.8 g/kg ideal body weight (IBW) for sedentary adults; higher in athletes	Protein is necessary for performance, preventing muscle breakdown and for workout recovery; protein sources from foods preferred over supplements	• Protein supplementation can worsen renal impairment • Avoid soy supplements if history of breast cancer • High-dose protein supplements can cause gastrointestinal distress
Branched-chain amino acids (BCAAs)	Dietary proteins should be obtained from food, per individual requirements	Benefit in certain disease states; evidence for benefit in athletes lacking	• May increase plasma ammonia levels
Glutamine	Not applicable	Benefit in certain disease states; evidence for benefit in athletes lacking	• Avoid in patients with seizure disorders or psychiatric illness • May increase plasma ammonia levels
L-Carnitine	Not applicable; 2 to 4 g/day used in studies	Evidence for benefit in athletes lacking	• Gastrointestinal effects (uncommon) • Fishy body odor
Creatine	20 g/day (or 0.3 g/kg) for 5 days followed by a maintenance dose of 2 or more g (or 0.03 g/kg) day (this is a typical regimen; other regimens also used)	May provide benefit for short, intense workouts by increasing muscle performance	• Gastrointestinal effects, primarily cramping • Weight gain attributable to water retention • Do not use in renal impairment • Avoid use in children
Carbohydrate-based "sport drinks"	Weight and duration dependent; see text	Carbohydrate and fluid are essential; inclusion of electrolytes controversial	• High glucose concentrations cause gastrointestinal upset and contribute to weight gain and dental caries • Excessive intake may cause hyponatremia

recovery, and decrease muscle breakdown from exercise. Stored fat is the energy source of last resort. The normal physiologic functions that are accelerated and often stressed during exercise all require numerous vitamins and minerals as cofactors to support their metabolic pathways. Many of these are now sold as supplements.

TOTAL CALORIES AND DIETARY COMPOSITION

Calorie requirements vary with age, size, sport, and the intensity of training. For the training athlete who exercises vigorously several hours per day, caloric

---

PATIENT CASES

MEGAN

Megan is a 53-year-old female. She has been active most of her life and jogs at least 5 days a week. She has noticed that she has gradually put on a few extra pounds over the past few years surrounding menopause, which occurred at age 51. Recently, she decided to participate in a week-long bicycle ride for charity. She has felt a little "achey" recently, in spite of increased stretching and an occasional yoga class.

Megan hopes that increasing vigorous exercise will help her to feel better in general and will help her to lose the extra weight. Although she has always been active, she has never considered herself an athlete and has never pursued the consistent training required to participate in a strenuous event such as a week-long bicycle ride. Megan will need increased endurance and strength to complete the long rides and make it up some of the impressive hills on the route. She would like some help in choosing the proper natural products to help her train for, and complete, the bicycle ride.

RON

Ron is a 32-year-old sales associate for a large company that manufactures computer modems. He spends a lot of time in his car and on airplanes and has gotten a little bigger around the middle over the past few years. He eats well on occasions, but also consumes a typical fast-food diet (hamburger, fries, and cola) four or five times a week. He has been maintaining a regular, moderate exercise and weight-training routine, but states he has not seen acceptable results. He wants to reduce his weight and increase his muscle definition.

Ron weighs 200 pounds and states he "looked good" at his college weight of 180 pounds. He believes proper exercise and nutrition will also decrease his risk of developing the health problems of his aging parents. Ron would like to add some nutritional supplements to increase his muscle mass and help compensate for poor eating habits.

requirements can be as much as 23 to 39 kcal per pound of body weight per day. A quick way to approximate intake is to multiply ideal body weight in pounds by 13 for a sedentary person, and by up to 18 for someone moderately to highly active.

Protein

Lean body mass is accrued by an increase in protein synthesis that exceeds the rate of breakdown. It requires the presence of sufficient protein intake. A healthy, sedentary adult requires about 0.8 g of protein/kg of ideal body weight (that is, about 0.36 g/pound).

There is evidence that protein supplementation following a strenuous workout hastens recovery due to a decrease in postexercise muscle breakdown. Two of the most commonly used sources of the protein in supplements are soy, a vegetarian product, and whey, produced from cow's milk.[1] They are fairly similar in many respects however studies show vegetable protein to be nutritionally equivalent to animal protein. Adding carbohydrates to protein supplements can provide additional benefit.[2,3]

Soy and Whey Protein Supplements

Soy protein may contribute to cholesterol reduction.[4,5] Whey contains several types of nutrients including carbohydrates, proteins, minerals, and immunoglobulins.[6] The bioactive peptides in whey, which are thought to be immune enhancing, can be altered or destroyed by processing methods, including temperature changes during production.[7,8] With several whey-based supplements on the market, it is difficult to determine the potency of their immunoglobulin components without knowing about the manufacturer and type of processing employed. However, the protein content should be listed on the labels for easy comparison.

Very few published studies are available on the effectiveness of whey for body-building. The few trials available show little to no evidence that whey works specifically to enhance body-building.[9,10] However, as with soy, whey is appropriate when used simply as a protein supplement to

Performance enhancers

ensure adequate protein intake during training. Soy has some added health benefits but many of those added health benefits require that the soy isoflavones be left intact during processing. Soy products with the isoflavones intact are usually labeled with isoflavone content.

To attain a well-balanced mix of amino acids, it is best to use protein from a variety of sources, with animal sources being the most complete in terms of amino acid content. Combining various vegetable protein sources to obtain all the necessary amino acids in the needed proportions on a daily basis becomes essential on a totally vegan diet. Whey has slightly more available protein than soy and may have immune-boosting benefits, particularly when it is processed in such a way as to maximize retention of its immunoglobulin components.

Safety Considerations/Drug Interactions

One concern with very high protein intake is the possibility of renal impairment. Several studies suggest that protein intake of up to 2.8 g/kg/day is well tolerated in athletes. These studies, as well as data on indigenous people with very high protein intake, indicate that the ceiling for human protein consumption is probably considerably higher.[11-13] However, caution must be used in certain patients. One large study of a group of women with mild renal impairment found that a high-protein diet put this group at risk for further deterioration of renal function.[14] More studies are required to determine whether there are any long-term effects of chronic protein loading in healthy athletes with no renal impairment. Protein should not be oversupplemented, especially in cases of even mildly compromised renal function.

Soy, taken in food form, is safe for all age groups. Although long-term soy intake may be protective against breast cancer in premenopausal women, it may not be wise to recommend soy products to women with a history of breast cancer, due to the possibility of estrogen-agonist effects.[15] Soy should be avoided in patients with soy allergy. Whey protein is well tolerated. High doses can cause gastrointestinal symptoms.

Branched-Chain Amino Acids

Emerging from studies of critical care, cirrhosis, and kidney dialysis patients is evidence that the branched-chain amino acids (BCAAs), leucine, isoleucine, and valine have beneficial anabolic effects in these patients. BCAAs are known to attenuate protein loss and improve nitrogen balance in cases of bed rest, liver failure, and kidney dialysis.[16-18] BCAAs appear to up-regulate tissue capacity to synthesize proteins and can be metabolized outside of the liver in skeletal muscle.[19,20]

Extrapolating from patients depleted by disease states, researchers postulated that there may be possible benefit from BCAA supplementation in athletes. Studies using rats suggest dietary BCAAs cause a sparing of glycogen stores during exercise.[21] While BCAAs have been reviewed elsewhere as not beneficial in enhancing athletic performance, there is one study demonstrating increased grip strength following 30 days of BCAA administration in a double-blind crossover study using healthy males and isocaloric supplements.[22]

There is evidence that previous estimates of both adult and child BCAA requirements were calculated as lower than actual needs, which reinforces the advisability that athletes consume sufficient quantities of protein containing BCAAs.[23,24] Eggs are a good source of dietary BCAAs. However, eggs may not contain the optimal balance of amino acids when consumed alone.[25] It is prudent to obtain dietary proteins from a variety of sources. BCAAs are also concentrated in fish, dairy, and germs of grains and are present in some "sport drinks."[26] Due to the lack of data showing benefit in BCAA supplementation for athletic performance at this time, it is best to recommend a variety of healthy food sources for BCAA content.

Safety Considerations/Drug Interactions

BCAA supplementation (i.e., not from food sources) poses the risk of increased plasma ammonia levels.[27] Short-term use of 60 g or less in persons with normal metabolic function, particularly normal hepatic function, may be safe.[28]

Glutamine

Glutamine is the most abundant amino acid in the human body, constituting 60% of the total intracellular amino acid pool.[29] Glutamine is used in protein synthesis, acts as a nitrogen transporter, is a substrate in ammonia production, and is an energy source for cells in the immune and gastrointestinal systems. Serum levels of glutamine have been shown to decrease with intense exercise and may remain low with overtraining.[30,31] Studies on the use of glutamine supplementation for performance enhancement have not shown improvements of short-term, high-intensity exercise, such as weightlifting or sprint cycling by trained athletes.[32,33] No other studies have been done on endurance performance or muscle growth in athletes.

Glutamine supplementation may be helpful for improved immune function in athletes, although this is speculative. Glutamine supplementation has become an accepted addition to nutritional formulas in trauma patients to improve nutritional and immunological status, and it has been shown to be useful in treating some types of diarrhea and AIDS-related wasting.[34]

Safety Considerations/Drug Interactions

Glutamine supplementation is well tolerated with no significant adverse effects in the majority of patients. In those with seizure disorders or psychiatric illness, it is best to avoid glutamine due to the risk of neurological effects. This dose-related effect can be caused by an increase in serum ammonia levels from glutamine metabolism.[35]

L-Carnitine

L-Carnitine is made from the amino acids lysine and methionine. It is produced by the liver, kidneys, and brain and can be obtained from food sources, particularly meat and dairy. L-Carnitine is involved in the transport of long-chain fatty acids across the mitochondrial membrane for energy production. Its presence is probably important in endurance exercise where fat comes into use as a source of energy.[36]

In spite of a dearth of pertinent research or direct supporting evidence, L-carnitine has become popular in performance-enhancing products as a potential ergogenic aid because of its role in the conversion of fat to energy. Research on the effects of 2 to 4 g/day of L-carnitine on performance is inconclusive.[37-39] Some reviewers have suggested that L-carnitine can attenuate the adverse effects of hypoxic training, speed recovery, and prevent cellular damage; however, substantial evidence is lacking on this as well.[40]

Safety Considerations/Drug Interactions

Adverse side effects when taken orally or intravenously are uncommon but can include nausea, vomiting, abdominal cramps, heartburn, gastritis, diarrhea, and a fishy body odor.

Creatine

After a discussion of the lack of benefit from BCAAs, except possibly from healthy food sources, and a lack of benefit at this time from glutamine or L-carnitine supplementation, the reader might be wondering if any of the amino acids or amino acid derivatives present in performance-enhancing products may be helpful.

Creatine appears to provide benefit for short, intense workouts by increasing muscle performance. Creatine is consumed in the diet and synthesized endogenously.[41] Creatine, along with adenosine triphosphate (ATP), is essential for cellular energy production. Possible mechanisms for the use of creatine in improving athletic performance include increased rates of phosphocreatine resynthesis, decreased accumulation of inorganic phosphate, and raised pH.[42]

There is also speculation that creatine ingestion stimulates the creatine kinase phosphocreatine system, which facilitates muscle relaxation.[43,44] In addition, creatine ingestion appears to stimulate muscle glycogen storage.[45] Vegetarian athletes may benefit more than nonvegetarians from creatine supplementation simply because their intake and stores are more likely to be low.[46]

Supplementation can increase the intramuscular creatine by 20%, which subsequently increases the phosphocreatine concentration and may allow for better energy availability during high-demand situations.[47,48] Skeletal muscle has a saturation point at which additional supplementation will not increase intracellular creatine levels. Creatine supplementation appears to temporarily down-regulate endogenous creatine production. After supplementation is discontinued, endogenous creatine synthesis and concentration typically return to baseline within 28 days.[47,49,50]

While creatine supplementation increases preexercise creatine and ATP stores, and appears to improve performance, exercise recovery may be decreased by creatine supplementation.[51] This may account for why, in one of the reviews of creatine use, weight-lifting capacity improved, but not cycling torque and sprinting, which do not involve the sort of intermittent bursts of effort that creatine loading most clearly improves.[52,53] However, another study concluded that 6 days of creatine supplementation did improve performance in sprint and agility tasks that simulated soccer. This improvement occurred even with an increased body mass that was assumed to be from an increase in water weight, secondary to the creatine supplementation.[54]

In spite of such discrepancies in the literature, the conclusion from a review article that considered more than 500 research studies seems to make sense, "Although not all studies report significant results, the preponderance of scientific evidence indicates that creatine supplementation appears to be a generally effective ergonomic aid for a variety of exercise tasks in a number of athletic and clinical populations."[55] It remains to be sorted out exactly what types of athletes and sports performances benefit from creatine supplementation. For example, there are contradictory study results concerning whether supplementation is of benefit to older men.[56-58]

Evidence for increases in lean body mass are less clear than the supporting data on ergonomic benefit. Most data suggest that the increase in body weight is not due to an increase in lean body mass, but rather due to an increase in water weight.[59-62]

The recommended dose for improving physical performance is typically given as a loading dose of 20 g/day (or 0.3 g/kg) for 5 days, followed by a maintenance dose of 2 or more g (or 0.03 g/kg/day).[48] However, other dosing regimens have been tried with favorable results.

Glucose seems to enhance the uptake of creatine by skeletal muscle and reduces the excretion of creatine; therefore, creatine supplements are typically dissolved in a carbohydrate solution for consumption.

SAFETY CONSIDERATIONS/DRUG INTERACTIONS

As for adverse effects, one review article aptly notes the lack of reported adverse events in spite of the widespread use of creatine in sports in the past 8 years.[63] Though side effects of creatine supplementation are not commonly seen, one study of male college athletes found that cramping occurred in 25% of the creatine users.[64] Other possible side effects include nausea and diarrhea.

Creatine often causes a weight gain of 0.5 to 1.6 kg that increases with prolonged supplementation and, as noted previously, is thought to be due to water retention.[62] People with kidney disease should avoid the use of creatine because of the risk of additional renal dysfunction.[65] Creatine is metabolized to creatinine, which is excreted renally.

Combining creatine with caffeine might increase the risk of serious adverse effects; therefore, this combination should be avoided.[66] Dehydration is possible and persons should be counseled to remain well hydrated, which is necessary anyway with any endurance-training program. It has been proposed that creatine increases protein synthesis or decreases protein catabolism, leading to increased muscle fibers.[48,67,68] Effects of creatine on growth and development are still unknown, and creatine should be avoided in children and adolescents.[48]

CARBOHYDRATES AND "SPORT DRINKS"

Carbohydrates are the most efficient fuel for direct energy production. Carbohydrates, in the form of glucose, can be pulled into cells directly from the bloodstream for immediate cellular conversion to energy. Carbohydrates can also be stored as glycogen in muscle and liver where they act as ready reserves of stored energy for strenuous exercise.

Duration and intensity of training determines the carbohydrate requirement. The glycogen stores can supply enough energy to last approximately 18 hours during normal levels of activity, but endurance exercise like biking up hills or marathon running can start depleting glycogen stores after 90 minutes.[69] Perhaps as a result of providing ready energy, carbohydrates taken during exercise appear to attenuate stress hormones such as cortisol and are one of the few interventions, out of many tried, that appear to limit the well-accepted phenomenon of exercise-induced immune suppression.[70]

An athlete may require up to 4.5 g of carbohydrates per pound of body weight per day, or 60% to 70% of total dietary calories from carbohydrates, whichever is greater. The range is anywhere from 6 to 10 g of carbohydrates per kilogram of body weight. An appropriate suggestion would be to start at the lower end of the range and titrate up, as required.

Consuming carbohydrate beverages before, during, and for at least 5 hours after a strenuous event can help maintain energy and mitigate muscle breakdown. In order to improve gastrointestinal tolerability, the carbohydrate content should be lower as one gets closer to the athletic event.[71]

A rehydration drink that is up to 6% carbohydrate, with low fructose if possible, would be a good option for both during and after exercise. Large amounts of fructose in these types of products can cause gastrointestinal distress. However, a mix of various forms of carbohydrates (e.g., glucose and fructose) may allow the use of varied intestinal transporters and help to achieve higher oxidation rates.[72] For endurance events lasting more than 1 hour, consuming one cup of a 6% to 8% carbohydrate drink every 15 or 20 minutes will supply 1 g of carbohydrate to the tissues every minute and will delay fatigue.[69]

Individual athletes should experiment to find out their tolerance for anything but low glycemic index carbohydrates and water in the 6 hours prior to events, because of the risk of inducing transient hypoglycemia.[73] However, high glycemic index carbohydrate foods, everybody's favorites, appear to be the most efficacious carbohydrates for consumption during events and for repleting glycogen stores in the first 24 hours after strenuous exercise.[74] Delaying carbohydrate intake after exercise greatly reduces the rate at which glycogen is resynthesized, so carbohydrate consumption should start as soon as the event ends. Adding protein to carbohydrate intake after exercise seems to further improve the rate of glycogen recovery.

SAFETY CONSIDERATIONS/DRUG INTERACTIONS

Concentrations of carbohydrates in fluids should be considered. Intestinal absorption of a 6% carbohydrate solution is similar to absorption of water alone. However, concentrations of glucose in the lumen approaching 10% can stimulate fluid secretions and cause gastrointestinal discomfort.[69] Excessive fluid intake, whether from plain water or sports drinks, can pose a risk for hyponatremia. Other concerns due solely to carbohydrate consumption include weight gain (from excessive intake) and dental caries.

FLUIDS, CALCIUM, MAGNESIUM, AND IRON SUPPLEMENTATION

Fluids should be consumed before, during, and after exercise. It is recommended to consume approximately two glasses of fluid 2 hours before exercise and at regular intervals during exercise. Flavored drinks which contain electrolytes may encourage athletes to drink more but there is controversy as to the necessity or role of electrolytes in sport beverages.[75,76] Inclusion of electrolytes in sport drinks has not been reported to cause problems. Replacement of electrolytes lost during exercise may be important, especially with significant losses which can occur in long, hot, endurance-type events. The important role of carbohydrate inclusion in sport beverages has already been discussed.

Antioxidants may help limit the oxidative damage that results from both exercise and from working out in polluted air. A supplement that does not contain excessive amounts of any agent can be used daily.

Calcium and magnesium are both essential for proper bone formation and reconstruction, as well as proper muscle functioning. Calcium is a mineral of great concern for young female athletes in particular. Intense training may cause very low body fat, which can lead to disruption of normal estrogen production and cessation of menstruation. This, in turn, increases the risk significantly for osteoporosis.[77] A deficiency in magnesium impairs athletic performance, and the average diet in developed countries is often magnesium-deficient. Dependent on dietary intake, these minerals may require supplementation.

Iron is an important mineral for athletes because it is a component of hemoglobin, which is responsible for carrying oxygen to muscle cells and removing carbon dioxide. Hemoglobin's efficiency in carrying oxygen profoundly affects performance. The positive performance effects accomplished via the outlawed and now obsolete practice of "blood doping" by transfusions are probably the reason iron supplementation for athletes has stimulated so much interest. Blood transfusions were performed on athletes to increase hemoglobin and red blood cell count.

Now erythropoietin supplementation and training at high altitudes can be employed to attain the same end, but iron supplementation would be a cheaper, easier, and safer alternative if the right dose and timing could be shown to improve blood oxygen carrying capacity. Unfortunately, iron supplementation in and of itself in nonanemic athletes has not been convincingly demonstrated to improve performance, but supplementation can reduce the risk of anemia in endurance athletes and others by helping to maintain iron stores. Studies using iron supplementation of 100 to 200 mg/day to try to improve exercise performance have shown conflicting results.[78]

Patient Discussion

Even though **Megan** wants to lose some weight, or fat to be precise, and **Ron** wants to trim his waistline, they should both take in additional calories on training days. They will need the extra calories to prevent muscle breakdown during exercise and recovery. Beyond the risks involved in muscle breakdown as an energy source, other consequences of inappropriate caloric restriction are chronic fatigue, sleep disturbances, reduced performance, impaired ability for intensive training, and increased vulnerability to injury.[79] Increasing muscle mass will help long term with the weight-loss process. Muscle tissue raises the basal metabolic rate and uses up more calories, even when at rest, than fat tissue.

Ron's caloric intake will need to be significantly higher than Megan's (135 pounds x 15 = 2025 kcal). His starting weight is higher and, as a male, he has a greater percentage of muscle mass. He is also younger and will have a higher resting metabolic rate. Ron weighs 200 pounds, but from weight/height charts and his previous weight, he thinks he should be about 180 pounds. Using this weight and an activity factor of 15 (180 pounds x 15 = 2700 kcal), an appropriate daily caloric intake can be calculated for days when he is not working out strenuously. For Ron's training days, an activity factor of 23 would be more appropriate. Ron has chosen a sport that requires perhaps the maximum amount of protein intake compared with most other sports, up to double that of sedentary consumption.

As a supplement, there are reasons why soy protein might be the best choice in Megan's case, and possibly for Ron too. Purely as a training aid, soy may be superior to whey protein in mitigating exercise-induced muscle degradation.[80] Megan will be training in an urban area with significant air pollution and there is evidence that soy is protective against oxidative stress.

To decide whether to recommend a protein supplement such as soy or whey to Ron or Megan, first calculate protein requirements based on ideal weight and activity level. Then counsel them to obtain most of the needed protein from food sources. Protein supplements can be used as needed to obtain optimal protein intake, especially surrounding strenuous or prolonged exercise. This will help limit muscle breakdown, speed recovery and support increases in muscle tissue. A protein supplement drink during and after workouts that contains carbohydrates may be of added benefit in terms of protecting protein stores in muscles from breakdown. Soy can be recommended for most patients.

If either athlete inquires concerning BCAA supplementation, both of them could be counseled that they would do just as well to simply include eggs, or at least egg whites, in their breakfasts on workout days. Glutamine and L-carnitine supplementation should not be recommended.

Ron will probably benefit from creatine supplementation while training, both to increase his strength and possibly to increase his muscle mass. Since Megan calculated that she has been eating a protein-deficient diet, she might benefit from creatine supplementation for bursts of strength, such as when she is weight training. Fluid retention is likely to be a concern in any weight-conscious person. Megan should be counseled on the importance of getting sufficient protein, either by adding more animal protein or appropriate combinations of vegetarian foods. Supplementation with specific amino acids is not necessary or useful as long as a variety of dietary proteins is consumed.

Megan should be advised to consume carbohydrate-containing beverages before, during, and for at least 5 hours after long, strenuous bicycle rides. A good recommendation would be a drink that is up to 6% carbohydrate and low in fructose. Ron can use these drinks if he is planning a heavy workout schedule. Over consumption is not recommended.

An antioxidant/multivitamin supplement that does not contain excessive amounts of any agent can be recommended for both Ron and Megan. These supplements do not contain significant amounts of calcium and magnesium, and these will need to be considered separately. Ron and Megan should both consume calcium and magnesium in the diet, but they probably should take supplements in addition. Megan, in particular, should be careful to calculate her daily calcium intake. The RDA for her age group is 1200 mg/day. Any required calcium supplementation should be taken in divided doses. Many calcium supplements also contain safe amounts of magnesium, though taking two at the same time probably decreases absorption of either one..

Iron supplementation may not be required in either patient, if daily intake is sufficient. If there is any doubt, many multivitamins contain suitable supplementation to avoid deficiency, such as 18-mg ferrous fumarate (in One-A-Day® Women's Formula, and others).

SELF-ASSESSMENT

1. **Which is TRUE concerning caloric requirements:**
   a. It is best to maintain a steady, constant caloric intake on sedentary days and during periods of intense physical training.
   b. Inadequate calorie consumption does not contribute to muscle wasting.
   c. A quick way to calculate approximate daily intake is to multiply your ideal weight by an activity factor.
   d. The most important contributor to Ron's weight gain is cortisol levels.

2. **What is the approximate daily protein requirement for healthy, sedentary adults?**
   a. 0.2 g of protein per kilogram of ideal body weight
   b. 0.2 g of protein per kilogram of actual body weight
   c. 0.8 g of protein per kilogram of ideal body weight
   d. 2.0 g of protein per kilogram of ideal body weight

3. Which is TRUE concerning soy as a protein supplement:
   a. It should be avoided in vegetarians.
   b. It can help control cholesterol levels.
   c. It has a higher percentage of useable protein than whey.
   d. It is safe, in any amount, in persons with renal impairment.

4. Branched-chain amino acid (BCAA) supplementation is popular in many performance-enhancing products. This is because:
   a. There are significant benefits to regular consumption of BCAAs.
   b. BCAA supplements are superior to amino acid consumption from healthy food sources.
   c. It is not possible for endurance athletes to consume the recommended amounts of BCAAs required in the diet.
   d. They are profitable.

5. Which of the following can increase serum ammonia levels, particularly with high-dose, chronic use?
   a. Glutamine
   b. BCAAs
   c. Soy
   d. a and b

6. L-Carnitine can be recommended to most athletes. Supplementation provides significant health benefits.
   a. True
   b. False

7. Which is TRUE concerning creatine supplementation:
   a. Benefits are primarily present during short, intense regimens.
   b. Creatine supplementation can provide significant increases in muscle mass.
   c. Creatine causes significant gastrointestinal effects at typical doses and care must be used during supplementation.
   d. Creatine is safe to recommend in children who participate in rigorous sports.

8. Carbohydrates and fluids should be consumed:
   a. Before strenuous activity
   b. During strenuous activity
   c. For at least 5 hours after strenuous activity
   d. All of the above

9. Rehydration drinks that contain greater than 2% carbohydrate content can cause gastrointestinal distress.
   a. True
   b. False

10. Female athletes, even at a young age, must calculate daily calcium intake to ensure adequate intake.
    a. True
    b. False

Answers: 1-c; 2-c; 3-b; 4-d; 5-d; 6-b; 7-a; 8-d; 9-b; 10-a

REFERENCES

1. Jeejeebhoy KN. Vegetable proteins: are they nutritionally equivalent to animal protein. *Eur J Gastroeneterol Hepatol.* 2000;12(1):1-2.

2. Rasmussen BB, Tipton KD, Miller SL, et al. An oral essential amino acid-carbohydrate supplement enhances muscle protein anabolism after resistance exercise. *J Appl Physiol.* 2000;88:386-92.

3. Miller SL, Tipton KD, Chinkes DL, et al. Independent and combined effects of amino acids and glucose after resistance exercise. *Med Sci Sports Exerc.* 2003;35:449-55.

4. Aoyama T, Fuki K, Nakamori T, et al. Effect of soy and milk whey isolates and their hydrolysates on weight reduction in genetically obese mice. *Biosci Biotechnol Biochem.* 2000;64(12):2594-600.

5. Lavigne C, Marette A, Jaques H. Cod and soy proteins compared with casein improved glucose tolerance and insulin sensitivity in rats. *Am J Physiol Endocrinol Metab.* 2000;278(3):E491-500.

6. Jellin JM. Whey. Natural Medicines Comprehensive Database, 2002.

7. Shah NP. Effects of milk-derived bioactives: an overview. *Br J Nutr.* 2000;84(Suppl 1):S2-S10.

8. Florisa R, Recio I, Berkhout B, et al. Antibacterial and antiviral effects of milk proteins and derivatives thereof. *Curr Pharm Des.* 2003;9(16):1257-75.

9. Agin D, Gallagher D, Wang J, et al. Effects of whey protein and resistance exercise on body cell mass, muscle strength, and quality of life in women with HIV. *AIDS.* 2001;15:2431-40.

10. Burke DG, Chilibeck PD, Davidson KS, et al. The effect of whey protein supplementation with and without creatine monohydrate combined with resistance training on lean tissue mass and muscle strength. *Int J Sport Nutr Exerc Metab.* 2001;11:349-64.

11. Poortmans JR, Dellalieux O. Do regular high protein diets have potential health risks on kidney function in athletes? *Int J Sport Nutr Exerc Metab.* 2000;10(1):28-38.

12. Manz F, Remer T, Decher-Spliethoff E, et al. Effects of a high protein intake on renal acid excretion in bodybuilders. *Z Ernahrungswiss.* 1995;34(1):10-5.

13. Gadspy P. The Intuit paradox. *Discover.* 2004;25(10):52.

14. Knight EL, Stampfer MJ, Hankinson SE, et al. The impact of protein intake on renal function decline in women with normal or mild renal insufficiency. *Ann Intern Med.* 2003;138:460.

15. Murkies A, Dalais FS, Briganti EM, et al. Phytoestrogens and breast cancer in postmenopausal women: a case control study. *Menopause.* 2000;7:289-96.

16. Stein TP, Schluter MD, Leskiw MJ, et al. Attenuation of the protein wasting associated with bedrest by branched-chain amino acids. *Nutri Rev.* 1999;62(1):33-8.

17. Mascarenthas R, Mobarhan S. New support for branched-chain amino acid supplementation in advanced hepatic failure. *Nutri Rev.* 2004;62(1):33-8.

18. Vuzelov E, Krivoshiev S, Ribarova F, et al. Plasma levels of branched chain amino acids in patients on regular hemodialysis before and after including a high-protein supplement in their diet. *Folia Med* (Plovdiv). 1999;19-22.

19. Kimball SR, Jefferson LS, Regulation of protein synthesis by branched-chain amino acids. *Curr Opin Clin Nutr Metab Care.* 2001;4(1):39-43.

20. Lui Z, Jahn LA, Long W, et al. Branched chain amino acids activate messenger ribonucleic acid translation regulatory proteins in human skeletal muscle, and glucocorticoids blunt this action. *J Clin Endocrinol Metab.* 2001;86(5):2136-43.

21. Shimomura Y, Murakami T, Nakai N, et al. Suppression of glycogen consumption during acute exercise by dietary branched-chain amino acids in rats. *J Nutr Sci Vitaminol.* 2000;46(2):71-7.

22. De Lorenzo A, Petroni ML, Masala S, et al. Effect of acute and chronic branched-chain amino acids on energy metabolism and muscle performance. *Diabetes Nutr Metab.* 2003;16(5-6):291-7.

23. Riazi R, Wykes LJ, Ball RO, et al. The total branched-chain amino acid requirement in young healthy adult men determined by indicator amino acid oxidation by use of L-[1-13C] phenylalanine. *J Nutr.* 2003;133(5):1383-9.

24. Mager DR, Wykes LJ, Ball RO, et al. Branched-chain amino acid requirements in school-aged children determined by indicator amino acid oxidation. *J Nutr.* 2003;133(11):3540-5.

25. Raizi R, Rafii M, Wykes LJ, et al. Valine may be the first limiting branched-chain amino acid in egg protein in men. *J Nutr.* 2003;133(11):3533-9.

26. Tipton KD, Ferrando AA, Phillips SM, et al. Postexercise net protein synthesis in human muscle from orally administered amino acids. *Am J Physiol Endocrinol Metab.* 1999;276:E628-34.

27. MacLean DA, Graham TE, Saltin B. Branched-chain amino acids augment ammonia metabolism while attenuating protein breakdown during exercise. *Am J Physiol.* 1994;267:E1010-22.

28. Jellin JM. Valine. Natural Medicines Comprehensive Database, 2005. Available at www.naturaldatabase.com. Accessed March 23, 2005.

29. Bergström JV Jr, Fürst P, Noree LO, et al. Intracellular free amino acid concentration in human muscle tissue. *J Appl Physiol.* 1974;36:693.

30. Antonio J, Street C. Glutamine: a potentially useful supplement for athletes. *Can J Appl Physiol.* 1999;24:1-14.

31. Halson SL, Lancaster GI, Jeukendrup AE, et al. Immunological responses to overreaching in cyclists. *Med Sci Sports Exerc.* 2003 May;35(5):854-61.

32. Rohde T, MacLean DA, Pedersen BK. Effect of glutamine supplementation on changes in the immune system induced by repeated exercise. *Med Sci Sports Exerc.* 1998;30:856-62.

33. Cynober LA. Do we have unrealistic expectations of the potential of immunonutrition? *Can J Appl Physiol.* 2001;26 Suppl:S36-44. Review.

34. Jones C, Palmer TE, Griffiths RD. Randomized clinical outcome study of critically ill patients given glutamine-supplemented enteral nutrition. *Nutrition.* 1999;15:108-15.

35. Ward E, Picton S, Reid U, et al. Oral glutamine in paediatric oncology patients: a dose finding study. *Eur J Clin Nutr.* 2003;57:31-6.

36. Kiens B, Roepstorff C. Utilization of long-chain fatty acids in human skeletal muscle during exercise. *Acta Physiol Scand.* 2003;178(4):391-6.

37. Kanter MM, Williams MH. Antioxidants, carnitine, and choline as putative egogenic aids. *Int J Sport Nutr.* 1995;5(Suppl):S120-31.

38. Heinonen OJ. Carnitine and physical exercise. *Sports Med.* 1996;22(2):109-32.

39. Colombani P, Wenk C, Kunz I, et al. Effects of L-carnitine supplementation on physical performance and energy metabolism of endurance-trained athletes: a double-blind crossover field study. *Eur J Appl Physiol Occup Physiol.* 1996;73(5):434-9.

40. Karlic H, Lohniger A. Supplementation of L-carnitine in athletes: does it make sense? *Nutrition.* 2004;20(7-8):709-15.

41. Persky AM, Brazeau GA. Clinical pharmacology of the dietary supplement creatine monohydrate. *Pharmacol Rev.* 2001;53:161-76.

42. Yquel RJ, Arsac LM, Thiaudiere E, et al. Effect of creatine supplementation on phosphocreatine resynthesis, inorganic phosphate accumulation and pH during intermittent maximal exercise. *J Sports Sci.* 2002;20(5):427-37.

43. Hespel P, Eijnde BO, Derave W, et al. Creatine supplementation: exploring the role of the creatine kinase/phosphocreatine system in human muscle. *Can J Appl Physiol.* 2001;26(Suppl):S79-102.

44. Shabert JK, Winslow C, Lacey JM, et al. Glutamine-antioxidant supplementation increases body cell mass in AIDS patients with weight loss: a randomized, double-blind controlled trial. *Nutrition.* 1999;15:860-4.

45. Van Loon LJ, Murphy R, Oosterlaar AM, et al. Creatine supplementation increases glycogen but not GLUT-4 expression in human skeletal muscle. *Clin Sci.* 2004;106(1):99-106.

46. Lukaszuk JM, Robertson RL, Arch JE, et al. Effect of creatine supplementation and a lacto-ovo-vegetarian diet on muscle creatine concentration. *Int J Sport Nutr Exerc Metab.* 2002;12(3):336-48.

47. Jellin JM. Creatine. Natural Medicines Comprehensive Database. 2002;423-5.

48. Racette SB. Creatine supplementation and athletic performance. *J Orthop Sports Phys Ther.* 2003;33(10):615-21.

49. Vandenberghe K, Goris M, Van Hecke P, et al. Long-term creatine intake is beneficial to muscle performance during resistance training. *J Appl Physiol.* 1997;83(6):2055-63.

50. McKenna MJ, Morton J, Selig SE, et al. Creatine supplementation increases muscle total creatine but not maximal intermittent exercise performance. *J Appl Physiol.* 1999;87(6):2244-52.

51. Kreis R, Kamber M, Koster M, et al. Creatine supplementation II: in vivo magnetic resonance spectroscopy. *Med Sci Sport Exerc.* 1999;31(12):1770-7.

52. Dempsy RL, Mazzone MF, Meurer LN. Does creatine supplementation improve strength? A meta-analysis. *J Fam Pract.* 2002;51(11):945-51.

53. Vandenberghe K, Van Hecke P, Van Leemputte M, et al. Phosphocreatine resynthesis is not affected by creatine loading. *Med Sci Sport Exerc.* 1999;31(2):236-42.

54. Cox G, Mujika I, Tumilty D, et al. Acute creatine supplementation and performance during a field test simulating match play in elite female soccer players. *Int J Sport Nutr Exerc Metab.* 2002;12(1):33-46.

55. Kreider RB, Ferreira M, Wilson M, et al. Effects of creatine supplementation on performance and training adaptation. *Mol Cell Biochem.* 2003;244(1-2):89-94.

56. Eijinde BO, Van Leemputte M, Goris M, et al. Effects of creatine supplementation and exercise training on fitness in men 55-75 yr old. *J Appl Physiol.* 2003;95(2):818-28.

57. Gotshalk LA, Volek JS, Staron RS, et al. Creatine supplementation improves muscular performance in older men. *Med Sci Sport Exerc.* 2002;34(3):537-43.

58. Rawson ES, Clarkson PM, Price TB, et al. Differential response of muscle phosphocreatine to creatine supplementation in young and old subjects. *Acta Physiol Scand.* 2002;174(1):57-65.

59. Hultman E, Soderlund K, Timmons JA, et al. Muscle creatine loading in men. *J Appl Physiol.* 1996;81:232-7.

60. Parise G, Mihic S, MacLennan D, et al. Effects of acute creatine monohydrate supplementation on leucine kinetics and mixed-muscle protein synthesis. *J Appl Physiol.* 2001;91:1041-7.

61. Potteiger JA, Webster MJ, Nickel GL, et al. The effects of buffer ingestion on metabolic factors related to distance running performance. *Eur J Appl Physiol Occup Physiol.* 1996;72:365-71.

62. Kutz MR, Gunter MJ. Creatine monohydrate supplementation on body weight and percent body fat. *J Strength Cond Res.* 2003;17(4):817-21.

63. Lemon PW. Dietary creatine supplementation and exercise performance: why inconsistent results? *Can J Appl Physiol.* 2002;27(6):663-81.

64. Juhn MS, O'Kane JW, Vinci DM. Oral creatine supplementation in male collegiate athletes: a survey of dosing habits and side effects. *J Am Diet Assoc.* 1999;99:593-5.

65. Juhn MS. Oral creatine supplementation. Separating fact from hype. *Phys Sportsmed.* 1999;27:47-50,53-54,56,61,89.

66. Persky AM, Brazeau GA, Hochhaus G. Pharmacokinetics of the dietary supplement creatine. *Clin Pharmacokinet.* 2003;42(6):557-74. Review.

67. Volek JS, Duncan ND, Mazzetti SA, et al. Performance and muscle fiber adaptations to creatine supplementation and heavy resistance training. *Med Sci Sports Exerc.* 1999;31:1147-56.

68. Willoughby DS, Rosene J. Effects of oral creatine and resistance training on myosin heavy chain expression. *Med Sci Sports Exerc.* 2001;33:1674-81.

69. Mahan KL. *Krause's Food, Nutrition & Diet Therapy.* S. Escott-Stump, ed. Philadelphia: Saunders; 2004.

70. Gleeson M, Lancaster GI, Bishop NC. Nutritional strategies to minimize exercise-induced immunosuppression in athletes. *Can J Appl Physiol.* 2001;26(Suppl):S23-35.

71. Coleman E. Carbohydrates—the master fuel. Bering JR, Steen SN, eds. In: *Nutrition for Sport and Exercise*, Gaithersburg, MD: Aspen Publishers; 1998.

72. Jeukendrup AE. Carbohydrate intake during exercise and performance. *Nutrition.* 2004;20(7-8):669-77.

73. Hargreaves M, Hawley JA, Jeukendrup A. Pre-exercise carbohydrate and fat ingestion: effects on metabolism and performance. *J Sports Sci.* 2004;22(1):31-8.

74. Walton P, Rhodes EC. Determinants of post-exercise glycogen synthesis during short-term recovery. *Sports Med.* 1997;33(3):117-44.

75. Latzka WA, Montain SJ. Water and electrolyte requirements for exercise. *Clin Sports Med.* 1999;18(3):513-24.

76. Luethkemeier MJ, Coles MG, Askew EW. Dietary sodium and plasma volume levels with exercise. *Sports Med.* 1997;23(5):279-86.

77. Manore MM. Dietary recommendations and athletic menstrual dysfunction. *Sports Med.* 2002;32(14):887-901.

78. Brutsaert TD, Hernandez-Cordero S, Rivera J, et al. Iron supplementation improves progressive fatigue resistance during dynamic knee extensor exercise in iron-depleted, nonanemic women. *Am J Clin Nutr.* 2003;77(2):441-8.

79. Healthnotes, Athletic Performance. 2004. Available at www.healthnotes.com. Accessed March 21, 2005.

80. Nikawa T, Ikemoto M, Sakai T. et al. Effects of a soy protein diet on exercise-induced muscle protein catabolism in rats. *Nutrition.* 2000. 18(6):490-5.

Index

Note: Italicized letters *f* and *t* following page numbers indicate figures and tables, respectively.

A

acetaminophen, 37, 39, 44
acetylcholine, 19, 104
acetylcholinesterase inhibitors, 5, 18, 19. *See also* huperzine A
ACOG. *See* American College of Obstetricians and Gynecologists
Actonel, 73, 79
adenosine triphosphate (ATP), 29, 145, 215, 216
African plum tree. *See* pygeum
Agency for Healthcare Research and Quality, 144
AHA. *See* American Heart Association
ALA. *See* alpha-linolenic acid
alcohol, 141*f*
    for cardiovascular disease, 136*t*, 139–142
    for cholesterol reduction, 141
    as prostate irritant, 68
allergic rhinitis, 29
allergy
    to aspirin, 200
    to echinacea, 177, 184
    to feverfew, 28
    to psyllium, 165
    to soy, 213
    to stevia, 165
allicin, 123, 144
alliin, 123
alliinase, 123, 144
alpha-amylase, 201
alpha-linolenic acid (ALA), 124, 142, 143
alpha-lipoic acid, 200
"alternative" medicine, 10
AltMedDex System, 6
aluminum, 179
Alzheimer's disease. *See also* dementia
    development of, 18
    diagnosis of, 15
    DMEA for, 201

    ginkgo biloba for, 16, 20
    huperzine A for, 19
    incidence of, 15
    patient case on, 17, 20
American College of Obstetricians and Gynecologists (ACOG), 85–86
American Diabetes Association, 159
American dwarf palm tree. *See* saw palmetto
American ginseng, 182
    for cold and flu, 174*t*, 182–183
    for diabetes, 161*t*, 166–167, 168, 183
American Heart Association (AHA), 117, 126, 138, 141–143
American Urological Association (AUA), 63, 65, 67, 68
amino acids, 120, 144, 213, 214. *See also specific amino acids*
amitriptyline, 39, 44
amlodipine, 140
ammonia, 214, 215
amylases, 201
amyloid beta proteins, 18
ancient natural products, 9
androgens, 54, 65
andrographis, 174*t*, 175–176
andrographolide, 175
androstenedione, 54
angina pectoris, 137
animal sterol, 121
antibacterial agents
    calcium and, 75
    development of, 10
    magnesium and, 31–32
    zinc and, 180
anticoagulant agents, 8. *See also* warfarin
    andrographis and, 176
    chondroitin and, 41
    dong quai as, 89
    feverfew and, 28
    omega-3 fatty acids and, 107, 144
    policosanol and, 123
    vitamin E and, 139

antidepressants, tricyclic, 99, 100, 102–104
anti-inflammatory properties
   of butterbur, 28
   of evening primrose oil, 90
   of feverfew, 27
   of ginkgo biloba, 17, 18
   of glucosamine and chrondroitin, 41
   of omega-3 fatty acids, 43, 106, 143
   of pygeum, 66
   of stinging nettle, 67
antioxidants. *See also* vitamin C; vitamin E
   for dementia, 18, 19
   for performance enhancement, 219, 221
antiplatelet agents, 8. *See also* aspirin
   alcohol as, 141
   andrographis and, 176
   feverfew and, 28
   garlic and, 124, 144
   ginkgo biloba and, 18
   omega-3 fatty acids and, 107, 126, 143, 144
   policosanol and, 122–123
   vitamin E and, 20, 139
antispasmodic properties
   of butterbur, 28
   of dong quai, 89
antithrombotic properties
   of garlic, 144
   of omega-3 fatty acids, 106
"apple" shape, 195
arachidonic acid, 125
arginine
   for cardiovascular disease, 137t, 148–150
   for erectile dysfunction, 52–54, 52t, 57
ArginMax, 53
arthritis. *See* osteoarthritis
The Arthritis Cure (Theodosakis), 39
ascorbic acid. *See* vitamin C
Asian ginseng (Panax ginseng), 182
   for cold and flu, 174t, 182–183
   for erectile dysfunction, 52t, 55–57, 55f
aspirin
   alcohol and, 142
   allergy to, 200
   feverfew and, 28
   ginkgo biloba and, 18
   for heart disease, 140
   omega-3 fatty acids and, 144
   from willow bark, 10
atherosclerosis, 138
   DHEA and, 55
   omega-3 fatty acids for, 143
athletic performance. *See* performance enhancement
α-tocopherol, 19, 138. *See also* vitamin E
atorvastatin, 30, 118, 127, 140, 147
ATP. *See* adenosine triphosphate
AUA. *See* American Urological Association
avocado and soybean unsaponifiables (ASU), 38t, 43, 45

B

BCAAs. *See* branched-chain amino acids
benign prostatic hyperplasia (BPH), 63–68
   diagnosis of, 63
   patient case on, 65, 68
   pygeum for, 64t, 66–68
   risk factors of, 63
   saw palmetto for, 64–66, 64t, 68
   severity of, 63
   stinging nettle for, 64t, 67, 68
   symptoms of, 63
   yohimbe and, 57
beta-adrenergic receptors, 197
beta-glucan, 117
beta-sitosterol, 120, 121
bile acids, 117
bioidentical hormone replacement therapy (BHRT), 5, 86t, 90–92
bipolar disorder, 38t, 42, 55, 105, 106
birth control pills, 102
bisphosphonate, 78
bitter melon, 160t, 161–162, 161f
bitter orange, 197–198
black cohosh, 85–88, 86f, 86t, 92
black tea, 196, 197
bleeding risk, 8
   alcohol and, 142
   andrographis and, 176
   dong quai and, 89
   evening primrose oil and, 90
   feverfew and, 28
   garlic and, 124, 145
   ginkgo biloba and, 18
   omega-3 fatty acids and, 44, 107, 126, 144
   policosanol and, 123
   vitamin E and, 139
   willow bark and, 200
blond psyllium. *See* psyllium
blood glucose, 159. *See also* diabetes

alpha-lipoic acid and, 200
American ginseng and, 166–167, 183
Asian ginseng and, 56, 183
bitter melon and, 162
caffeine and, 196
cassia cinnamon and, 164
chromium and, 162
exercise and, 159
glucosamine and, 41
gymnema and, 164
monitoring, 163, 168
prickly pear cactus and, 165–166
psyllium and, 164
"sport drinks" and, 218
stinging nettle and, 67
and weight loss, 194
blood pressure
andrographis and, 176
arginine and, 54, 150
Asian ginseng and, 56, 183
caffeine and, 196
calcium and, 73
coenzyme Q10 and, 146
ephedra and, 197
garlic and, 123, 144, 145
hawthorn and, 147
high
   with diabetes, 163
   patient case on, 139–140, 150–151
   as risk factor for heart disease, 137
omega-3 fatty acids and, 107, 124, 143
yohimbe and, 57
blood transfusion, 219
Boiron, 183
bone density, 71, 73, 77–79
BPH. *See* benign prostatic hyperplasia
branched-chain amino acids (BCAAs), 210*t*, 213–214, 220
Bratman, Steven, 6
breast cancer
   black cohosh and, 87
   soy and, 77, 89, 121, 213
butterbur, 26*t*, 28–29, 28*f*

C

caffeine
   American ginseng and, 167
   for cardiovascular disease, 197
   with creatine, 217
   for migraine, 25
   for weight loss, 194, 194*t*, 196–197

calcium, 74*f*
   for osteoporosis, 72–75, 72*t*, 78, 79, 219
   for performance enhancement, 219, 221
   for weight loss, 199
calcium carbonate, 74
calcium citrate, 74
calorie requirements, 211–212
capsaicin, 38*t*, 42–43
carbamazepine, 76, 105
Carb Blocker, 201
Carb Intercept, 194*t*, 201
carbohydrates, 209, 212, 217–218, 221
cardiovascular disease, 135–151. *See also* blood pressure; cholesterol
   alcohol/wine for, 136*t*, 139–142
   arginine for, 137*t*, 148–150
   bioidentical hormone replacement therapy and, 91, 92
   caffeine for, 197
   carnitine for, 137*t*, 150
   coenzyme Q10 for, 30, 136*t*, 145–147
   death rates of, 135, 138, 142
   evening primrose oil for, 90
   exercise for, 79
   garlic for, 136*t*, 144–145
   ginseng use in, 56, 183
   hawthorn for, 136*t*, 147–148
   magnesium use in, 31
   myth about, 135
   omega-3 fatty acids for, 43, 44, 124, 127, 136*t*, 142–144
   patient case on, 139–140, 150–151
   risk factors of, 115, 137
   soy for, 77, 89, 120
   vitamin E for, 136*t*, 138–139
carnitine
   for cardiovascular disease, 137*t*, 150
   for performance enhancement, 210*t*, 215, 220
   for weight loss, 198
carvedilol, 140, 151
cassia cinnamon, 160*t*, 163–164, 167
cat's claw, 11
cayenne, 42*f*
celecoxib, 40, 41
Cenestin, 5, 5*f*
Centers for Disease Control and Prevention, 125
Ceylon cinnamon, 163
cGMP. *See* cyclic guanosine monophosphate
chemistry, 9–10

chemotherapy, 4, 30, 147
chest pain. *See* angina pectoris
children
    echinacea use in, 177
    fish consumption of, 44, 144
    natural products for, 11
    Prozac use in, 101
China (ancient), 9
Chinese cinnamon. *See* cassia cinnamon
Chinese club moss, 5, 18
chitin, 39
chitosan, 201
chloral, 10
chloroform, 10
cholecalciferol. *See* vitamin D3
cholesterol. *See also* HDL cholesterol; LDL cholesterol
    as animal sterol, 121
    as risk factor for heart disease, 137
cholesterol reduction, 115–128
    alcohol for, 141
    cassia cinnamon for, 164
    garlic for, 116t, 123–124, 144
    glucomannan for, 200
    oats for, 116t, 117–119, 126–128
    omega-3 fatty acids for, 116t, 124–127
    patient case on, 117–118, 126–128
    plant sterols for, 116t, 121–122, 127, 128
    policosanol for, 11, 116t, 122–123
    psyllium for, 116t, 119
    soy for, 77, 89, 116t, 119–121, 212
    and weight loss, 194
chondrocytes, 40
chondroitin, 38t, 39–41, 44
chromium
    for diabetes, 160t, 162–163, 167, 199
    for weight loss, 199
chromium picolinate, 161, 162
cinchona, 9
cinnamon, 163. *See also* cassia cinnamon
ciprofloxacin, 31–32
CitriMax, 198
clopidogrel, 28
clozapine, 78
CNS depressants, 67
Cochrane Database of Systematic Reviews, 7
cocoa, 196
cod liver oil, 126
coenzyme Q10 (CoQ10)
    for cardiovascular disease, 30, 136t, 145–147

    for migraine, 26t, 30–32
cola nut, 196
cold, 173–184
    andrographis for, 174t, 175–176
    echinacea for, 174t, 176–178, 184
    elderberry for, 174t, 180–181, 184
    ginseng for, 174t, 182–183
    Oscillococcinum for, 174t, 183–184
    patient case on, 175, 184
    vitamin C for, 174t, 178–179
    zinc for, 174t, 179–180, 184
combination products. *See also* multiagent supplements
    for benign prostatic hyperplasia, 64, 67, 68
    for erectile dysfunction, 53
    for migraine, 32
    for osteoarthritis, 39–41, 44
    for osteoporosis, 74, 75, 78, 79
    and polypharmacy, 9
    sales of, 2
    use of, 11
    for weight loss, 196
"complementary" medicine, 10
ConsumerLab, 7, 42, 66, 122, 123, 126
contaminants, 7, 44, 125, 126, 144
coral calcium, 73, 74
coronary heart disease, 137. *See also* cholesterol
    omega-3 fatty acids for, 126
    risk factors of, 137
    soy for, 120
    vitamin E in prevention of, 138
creatine, 209, 210t, 215–217
creatine kinase, 216
cyclic guanosine monophosphate (cGMP), 51–52
cyclobenzaprine, 78
cyclosporine, 56, 102, 183
CYP 450 1A2, 78, 102
CYP 450 3A4, 52t, 55, 102, 124, 178
CYP 450 2C9, 78, 102
cysteine, 162

D

daidzen, 77
damiana, 58
DASH. *See* Dietary Approaches to Stop Hypertension
databases, on natural products, 6–7
degenerative joint disease. *See* osteoarthritis

dehydroepiandrosterone. *See* DHEA
dehydrotestosterone, 65
dementia, 15–20. *See also* Alzheimer's disease
    causes of, 15, 18
    ginkgo biloba for, 16–18, 16*t*, 20
    huperzine A for, 5, 16*t*, 18–19
    patient case on, 17, 20
    vitamin E for, 16*t*, 19–20
depression, 97–108
    incidence of, 99
    inositol for, 98*t*, 104–105, 108
    omega-3 fatty acids for, 98*t*, 105–108
    patient case on, 99, 108
    SAMe for, 98*t*, 102–104, 108
    St. John's wort for, 98*t*, 100–102, 108
Dexatrim, 194*t*, 197, 198, 200
DHA. *See* docosahexaenoic acid
DHEA, 52*t*, 54–55, 54*f*, 57
diabetes, 159–168. *See also* blood glucose
    American ginseng for, 161*t*, 166–168, 183
    bitter melon for, 160*t*, 161–162
    cassia cinnamon for, 160*t*, 163–164, 167
    chromium for, 160*t*, 162–163, 167, 199
    DHEA and, 55
    flu prevention in, 175, 184
    glucosamine and, 41
    gymnema for, 160*t*, 164
    with high blood pressure, 163
    with high cholesterol, 118, 128
    patient case on, 163, 167–168
    prickly pear cactus for, 161*t*, 165–168
    psyllium for, 160*t*, 164–165, 167
    as risk factor for heart disease, 137
    risk factors of, 159
    stevia for, 161*t*, 165
diazepam, 78
Dietary Approaches to Stop Hypertension (DASH), 199
dietary composition, 211–212
Dietary Supplement Health and Education Act (DSHEA), 2–3
dietary supplement industry, 1
Dietary Supplement Verification Program (DSVP), 7
diethylether, 10
digoxin, 102, 136*t*, 140, 148
2-dimethylaminoethanol (DMEA), 201
Dioscorides, 9
disulfiram, 142
diuretics, 180, 200

divalproex sodium, 27
docosahexaenoic acid (DHA), 105–106, 124, 142
    for cardiovascular disease, 143
    for cholesterol reduction, 125, 126
    for depression, 107
    foods containing, 125*t*
donepezil, 17, 20
dong quai, 86*t*, 87, 89, 92
dopamine, 103, 162
doxorubicin, 30, 147
Dr. Phil's Shape Up!, 195
drug interactions, 8. *See also specific drugs*
DSHEA. *See* Dietary Supplement Health and Education Act
DSVP. *See* Dietary Supplement Verification Program

E

echinacea, 174*t*, 176–178, 176*f*, 184
EFAs. *See* essential fatty acids
eicosapentaenoic acid (EPA), 105–106, 124, 142
    for cardiovascular disease, 143
    for cholesterol reduction, 125, 126
    for depression, 107
    foods containing, 125*t*
elderberry, 174*t*, 180–181, 181*f*, 184
electrolytes, 218–219
endometrial cancer, 92
endothelium-derived relaxing factor (EDRF), 148
environment, 11, 120
ephedra, 197
ephedra-like stimulants, 194, 194*t*, 197–198
erectile dysfunction (ED), 51–58
    arginine for, 52–54, 52*t*, 57
    Asian ginseng for, 52*t*, 55–57
    causes of, 51, 57
    DHEA for, 52*t*, 54–55, 57
    incidence of, 51
    patient case on, 53, 57–58
    saw palmetto and, 66
    yohimbe for, 52*t*, 56–58
erythropoietin, 219
essential fatty acids (EFAs), 124. *See also* omega-3 fatty acids
Ester-C, 178
Estrace, 90
estradiol, 5, 85, 90, 91
estriol, 90

estrogen
in bioidentical hormone replacement therapy, 5, 90–92
black cohosh and, 86–87
body fat and, 219
DHEA as precursor of, 54, 55
ipriflavone and, 78
in Premarin, 4–5, 90, 91
soy and, 76, 77, 88, 89, 121, 213
estrone, 90
"ether," 10
evening primrose oil, 86t, 89–90
Excedrin, 25
exercise
arginine and, 149
and blood glucose, 159
for cardiovascular health, 79, 139
carnitine and, 150
for osteoarthritis, 45
for weight loss, 202
ezetimibe, 116t, 122

F

Facts and Comparisons, 6
fasting plasma glucose (FPG), 159
FDA. See Food and Drug Administration
fenugreek seed, 200
fertilizers, 11
feverfew, 26–28, 26t, 27f, 32
fiber, 117, 119, 120, 128, 160, 164, 199, 200
finasteride, 65, 66
fish oil, 124f
for cardiovascular disease, 43, 44, 124, 127, 136t, 142–144
for cholesterol reduction, 116t, 124–127
for depression, 98t, 105–108
foods containing, 125t
for osteoarthritis, 38t, 43–45, 106
Fleming, Sir Alexander, 10
flu, 173–184
andrographis for, 174t, 175–176
echinacea for, 174t, 176–178, 184
elderberry for, 174t, 180–181, 184
ginseng for, 174t, 182–183
Oscillococcinum for, 174t, 183–184
patient case on, 175, 184
vitamin C for, 174t, 178–179
zinc for, 174t, 179–180, 184
fluids, 218–219
fluoxetine, 39, 99, 102

Food and Drug Administration (FDA), 3, 44, 117, 119–120, 142, 144, 165, 197
FPG. See fasting plasma glucose
free radicals, 18, 19
"French Paradox," 141
fructose, 121, 218
furosemide, 140

G

gamma-amino-butyric acid (GABA), 100
γ-linolenic acid, 89–90
garcinia, 198
garlic, 145f
for cardiovascular disease, 136t, 144–145
for cholesterol reduction, 116t, 123–124, 144
gastrointestinal side effects
of American ginseng, 167
of andrographis, 176
of butterbur, 29
of calcium, 75, 199
of carnitine, 150, 198, 215
of coenzyme Q10, 30, 146
of creatine, 217
of echinacea, 177
of elderberry, 181
of garlic, 124, 145
of ginkgo biloba, 18
of hawthorn, 148
of huperzine A, 19
of inositol, 105
of magnesium, 31
of oats, 119
of omega-3 fatty acids, 107, 126
of plant sterols, 122
of prickly pear cactus, 166
of psyllium, 165
of pygeum, 67
of SAMe, 104
of saw palmetto, 66
of soy, 213
of "sport drinks," 218
of St. John's wort, 101
of stevia, 165
of stinging nettle, 67
of vitamin C, 179
of vitamin E, 20
of willow bark, 200
of yohimbe, 57
of zinc, 180

gemfibrozil, 30, 136t, 147
ginger, 200
ginkgo biloba, 15f, 16–18, 16t, 20
ginkgolides, 17, 18
ginseng. *See also* American ginseng; Asian ginseng; Siberian ginseng
   for cold and flu, 174t, 182–183
   toxins in, 11
glucomannan, 200
glucosamine, 5, 38t, 39–41, 40f, 44
Glucosamine–Chondroitin Arthritis Intervention Trial, 40
glutamate, 100
glutamic acid, 162
glutamine, 210t, 214–215, 220
glyburide, 162
glycemic index, 218
glycine, 162
glycogen, 214, 217, 218
glycosaminoglycans, 39–40
Good Manufacturing Practices (GMPs), 7, 75
grape seed extract, 201
Greece (ancient), 9
green tea, 196, 197
guarana, 196
"gurmar." *See* gymnema
gymnema, 160t, 164

H

haloperidol, 78
Harkness, Richard, 6
hawthorn, 136t, 147–148
HCA. *See* hydroxycitric acid
HDL cholesterol
   alcohol and, 141
   and cardiovascular disease, 115
   DHEA and, 55
   plant sterols and, 121
   policosanol and, 122
headaches. *See also* migraine
   chromium and, 162
   ginseng and, 56, 183
   hawthorn and, 148
   prickly pear cactus and, 166
   SAMe and, 104
   saw palmetto and, 66
   stevia and, 165
   vitamin C and, 179
   yohimbe and, 57
   zinc and, 180

heart attack(s), 137
   family history of, 139, 150
   multiple, 140, 151
   vitamin E in prevention of, 138
HeartBar, 149
heart surgery, 140
hemoglobin, 219
hemorrhagic stroke, 138–139, 142
HMG-CoA reductase inhibitors, 115, 117, 142
hormone replacement therapy, 4–5. *See also* bioidentical hormone replacement therapy
horny goat weed, 58
hot flashes, 4, 85–88, 91, 92
huperzine A, 5, 16t, 18–19, 19f
hydroxychalcone, 163
hydroxycitric acid (HCA), 198
Hydroxycut, 194t, 197–200
hypericin, 101
hypericum. *See* St. John's wort
hypermagnesemia, 31
hypertension. *See* blood pressure, high

I

ibuprofen, 5, 37, 39
imipramine, 103
immunostimulant properties
   of andrographis, 175, 176
   of caffeine, 197
   of echinacea, 177–178
   of elderberry, 181
   of ginseng, 182, 183
   of whey, 213
impaired glucose tolerance (IGT), 159
"incremental glycemia," 167
indinavir, 179
influenza. *See* flu
inositol, 98t, 104–105, 108
insomnia
   chromium and, 162
   DHEA and, 55
   echinacea and, 177
   ginseng and, 52t, 56, 182, 183
   SAMe and, 104
   yohimbe and, 57
Institute of Medicine, 179, 180
"integrative" medicine, 10
international normalized ratio (INR), 18, 20
Inuit diet, 124, 142

ipriflavone, 72t, 77–79
iron, 219, 221
isoflavones, 76, 77, 213
isoleucine, 213
isopetasin, 28

J

J-curve, 141
joint damage/pain. See osteoarthritis

K

kidney bean extract, 201
knee pain, 39, 41

L

lactation
    feverfew during, 28
    fish consumption during, 44, 144
    inositol during, 105
    natural products during, 11
L-arginine. See arginine
L-carnitine. See carnitine
LDL cholesterol
    and cardiovascular disease, 115
    cassia cinnamon and, 164
    oats and, 117
    patient case on, 117, 118, 126–128
    plant sterols and, 121
    policosanol and, 122
    psyllium and, 119
    soy and, 120
    vitamin E and, 138
L-dopamine. See dopamine
leucine, 213
leukotriene synthesis, 28
levodopa, 104
licorice, 200
linolenic acid, 105. See also alpha-linolenic acid
γ-linolenic acid, 89–90
lithium, 105
liver toxicity, 8, 87–88
lovastatin, 30, 136t, 147
L-tyrosine. See tyrosine
lycopene, 63, 64, 68
lymphocytopenia, 78
lysine, 150, 198, 215

M

magnesium
    with calcium, 74
    for migraine, 25, 26t, 31–32
    for performance enhancement, 219, 221
MAGTAB, 198
mania, 55. See also bipolar disorder
margarine, 121
mate, 196
McGraw, Phil, 195
medroxyprogesterone acetate, 91
memantine, 20
memory loss. See also Alzheimer's disease; dementia
    statin-induced, 115–117
menopause, 85–92
    bioidentical hormone replacement therapy for, 5, 86t, 90–92
    black cohosh for, 85–88, 86t, 92
    dong quai for, 86t, 87, 89, 92
    evening primrose oil for, 86t, 89–90
    ipriflavone for, 77
    patient case on, 87, 92
    soy for, 4, 76, 86t, 88–89, 92
menstrual pain, 90
mercury, 44, 125, 126, 144
Metamucil, 119, 164
methionine, 102, 150, 198, 215
methylmercury, 44, 144
metoprolol, 30, 147
mevinolin, 117
Micromedex Health Care Series Databases, 6
midazolam, 178
midsummer daisy. See feverfew
migraine, 25–32
    butterbur for, 26t, 28–29
    caffeine for, 25
    coenzyme Q10 for, 26t, 30–32
    feverfew for, 26–28, 26t, 32
    incidence of, 25
    magnesium for, 25, 26t, 31–32
    patient case on, 27, 32
    riboflavin for, 26t, 29–30, 32
    symptoms of, 25
    triggers of, 32
    World Health Organization on, 25
"miracle" calcium, 73, 74
mirtazapine, 108
monoamine oxidase inhibitors, 56, 57, 100, 102, 167, 183

mood disturbance
  chromium and, 162
  in menopause, 86, 87
morphine, 9
multiagent supplements, 1–2. *See also* combination products
multivitamins, 11, 221
myocardial infarction. *See* heart attack

N

National Academy of Sciences, 76
National Cholesterol Education Program (NCEP), 115, 121
National Institutes of Health (NIH), 6, 40
natural, meaning of, 4–5, 19
Natural Medicines Comprehensive Database, 6
The Natural Pharmacist Natural Medicine Encyclopedia (Bratman and Harkness), 6
natural products. *See also specific products*
  claims made by, 3
  decision-making on, 4
  definition of, 2–3
  dosage of, 8
  efficacy of, 3, 4, 8
  and environment, 11
  history of, 9–10
  interest in, 2
  lack of regulation of, 3
  patients using, 8–11
  quality of, 7
  reputable resources on, 6–7
  safety of, 3
  selecting, 7
  top ten, 2t
  toxicity of, 4, 8
NCEP. *See* National Cholesterol Education Program
New York Heart Association (NYHA), 146–148, 151
niacin, 117
nicotinic acid, 162
night sweats, 85, 87
NIH. *See* National Institutes of Health
nitric oxide (NO), 51–53, 148–149
nonsteroidal anti-inflammatory drugs (NSAIDs), 5, 37, 41–44, 162
norepinephrine, 104, 162, 197
NSF International, 7

O

OA. *See* osteoarthritis
oatmeal, 117
oats, 116t, 117–119, 119f, 126–128
obesity, 195
Office of Dietary Supplements, 6
olanzapine, 78
oligomeric proanthocyanidins (OPCs), 147
omega-3 fatty acids, 124f
  for cardiovascular disease, 43, 44, 124, 127, 136t, 142–144
  for cholesterol reduction, 116t, 124–127
  for depression, 98t, 105–108
  foods containing, 125t
  for osteoarthritis, 38t, 43–45, 106
omega-6 fatty acids, 90, 124, 125
omeprazole, 16t, 18
opium, 9
oral contraceptives, 102
orange juice, 121
organic natural products, 11
Oscillococcinum, 174t, 183–184
osteoarthritis (OA), 5, 37–45
  avocado and soybean unsaponifiables for, 38t, 43, 45
  capsaicin for, 38t, 42–43
  glucosamine and chondroitin for, 38t, 39–41, 44
  omega-3 fatty acids for, 38t, 43–45, 106
  patient case on, 39, 44–45
  SAMe for, 38t, 41–42, 44, 103
  symptoms of, 37
osteomalacia, 76
osteopenia, 71
osteoporosis, 71–79
  calcium for, 72–75, 72t, 78, 79, 219
  diagnosis of, 71
  incidence of, 71–73
  ipriflavone for, 72t, 77–79
  patient case on, 73, 78–79
  prevention of, 72–73, 79
  soy for, 72t, 76–77
  symptoms of, 71
  vitamin D for, 72t, 75–79
"oxyflavone," 201

P

Pacific yew tree, 11
paclitaxel, 11
Panax ginseng. *See* Asian ginseng

Panax quinquefolius. See American ginseng
Paracelsus, 9
Parkinson's disease, 104
paroxetine, 102
parthenium. *See* feverfew
parthenolide, 27
Pauling, Linus, 178
PDE5. *See* phosphodiesterase type 5
"pear" shape, 195
pectin, 166
pemphigus vulgaris, 178
penicillin, 10
performance enhancement, 209–221
    branched-chain amino acids for, 210t, 213–214, 220
    carnitine for, 210t, 215, 220
    creatine for, 210t, 215–217
    glutamine for, 210t, 214–215, 220
    patient case on, 211, 219–221
    soy and whey protein supplements for, 210t, 212–213, 220
    "sport drinks" for, 210t, 217–219
pesticides, 11
petasin, 28
P-glycoprotein, 102
pharmacology, 9–10
Phaseolamin 2250 (Phase 2), 201
phenelzine, 56, 183
phenobarbital, 76
phenytoin, 76
phosphatidylinositol cycle, 104–105
phosphocreatine system, 215, 216
phosphodiesterase inhibitors, 52, 58
phosphodiesterase type 5 (PDE5), 52, 58
phospholipids, 104, 124
phytoestrogens, 76, 88, 89
phytosterols. *See* plant sterols
"P-insulin," 161
plant estrogens. *See* phytoestrogens
"plant insulin," 161
plant products. *See* natural products
plant sterols, 116t, 120–122, 127, 128
Pliny the Elder, 9
Plutarch, 26
policosanol, 11, 116t, 122–123
"polypeptide P," 161
polypharmacy, 9
"postfeverfew syndrome," 28
postprandial plasma glucose (PPG), 159, 164, 165, 167
potassium, 199–200
pravastatin, 30, 147

pregnancy
    American ginseng during, 167
    feverfew during, 28
    fish consumption during, 44, 125, 144
    inositol during, 105
    magnesium during, 32
    natural products during, 11
    riboflavin during, 30, 32
Premarin, 4–5, 90, 91
prickly pear cactus, 161t, 165–168, 166f
progesterone, 91
progestin, 91, 92
Promensil, 88
Prometrium, 90
propranolol, 27, 30, 78, 147
prostaglandin synthesis, 65
prostate cancer, 38t, 41, 63
prostate-specific antigen (PSA), 66
protease inhibitors, 102
protein, 120, 125, 209, 212, 218, 220, 221
protein supplements, 210t, 212–213, 220
proteoglycans, 39, 40
Prozac. *See* fluoxetine
pseudoephedrine, 197
psyllium
    for cholesterol reduction, 116t, 119
    for diabetes, 160t, 164–165, 167
pycnogenol, 53
pygeum, 64t, 66–68, 67f
pyrrolizidine alkaloids, 29

Q

quercetin, 201
quinine, 9
quinolone antibiotics, 31–32, 180
quinquefolans, 167

R

ramipril, 140, 151
red clover, 77, 88
red wine, 141
red yeast rice products, 117
rehydration drink, 218
Remifemin, 87
renal impairment, 213
The Review of Natural Products, 6
rheumatoid arthritis, 90
riboflavin, 26t, 29–30, 32
rickets, 76

rifampin, 76
Rome (ancient), 9

S

S-adenosylmethionine (SAMe), 103*f*
    for depression, 98*t*, 102–104, 108
    for osteoarthritis, 38*t*, 41–42, 44, 103
salicin, 10
Sambucel, 181
saw palmetto, 64–66, 64*t*, 65*f*, 68
schizophrenia, 105
seizures, 18
selective serotonin reuptake inhibitors, 101, 102, 108
selegiline, 20
semisynthetic products, 4
serotonin, 41, 103, 104, 106, 162
serotonin syndrome, 42, 102, 104
sertraline, 101, 102
The Sexy Years (Somers), 90
Siberian ginseng, 182, 200
sildenafil, 52, 53, 150
simvastatin, 30, 102, 136*t*, 147
sitosterolemia, 122
Sjogren's syndrome, 177
skin reactions
    to capsaicin, 43
    to echinacea, 177
    to ginkgo biloba, 18
    to ginseng, 167, 183
    to hawthorn, 148
    to St. John's wort, 101
    to stinging nettle, 67
Somers, Suzanne, 90
soy, 88*f*
    for cholesterol reduction, 77, 89, 116*t*, 119–121, 212
    foods containing, 120*t*
    for menopause, 4, 76, 86*t*, 88–89, 92
    for osteoporosis, 72*t*, 76–77
soy protein supplements, 210*t*, 212–213, 220
Spanish flu, 183
spironolactone, 140, 151
"sport drinks," 210*t*, 217–219
St. John's wort, 98*t*, 100–102, 100*f*, 108
Staphylococcus aureus, 10
statin-induced memory loss, 115–117
sterols, 120, 121. *See also* plant sterols
stevia, 161*t*, 165
steviol, 165

stevioside, 165
stinging nettle, 64*t*, 67, 68
substance P, 42, 43
sumatriptan, 27, 39
Sumerians, 9
synephrine, 194*t*, 197–198
synthetic products, 4, 5

T

tadalafil, 58
tamoxifen, 87
tea, 196, 197, 199
testosterone, 54, 65, 66
tetracycline, 180
Theodosakis, Jason, 39
theophylline, 78
therapeutic failure, 8
"Therapeutic Lifestyle Changes" (TLC), 126–127
Therapeutic Research Faculty, 6
α-tocopherol, 19, 138. *See also* vitamin E
toxicity, 4, 8
triazolam, 55
tricyclic antidepressants, 99, 100, 102–104
triptan agents, 25
T-score, 71, 73, 79, 87
type 2 diabetes. *See* diabetes
tyrosine, 201

U

ubiquinone. *See* coenzyme Q10
unipolar depression, 105, 106
urticaria, 176
U.S. Pharmacopeia (USP), 7, 10

V

valine, 213
valproate, 105
vanadium, 200
vasodilatory properties
    of arginine, 148–150
    of dong quai, 89
    of omega-3 fatty acids, 106, 143
venlafaxine, 108
verapamil, 78
versican, 41
Viagra. *See* sildenafil
vitamin B2. *See* riboflavin
vitamin C, 174*t*, 178–179

vitamin D, 72t, 74–76, 78, 79
vitamin D3, 75
vitamin E
    for cardiovascular disease, 136t, 138–139
    for dementia, 16t, 19–20
    dietary sources of, 138

W

warfarin
    American ginseng and, 167, 183
    Asian ginseng and, 56, 183
    chondroitin and, 41
    coenzyme Q10 and, 31, 147
    dong quai and, 89
    feverfew and, 28
    ginkgo biloba and, 18, 20
    ipriflavone and, 78
    omega-3 fatty acids and, 144
    St. John's wort and, 102
    stinging nettle and, 67
    vitamin C and, 179
    vitamin E and, 20
weight loss, 193–202, 196f
    caffeine for, 194, 194t, 196–197
    Carb Intercept for, 194t, 201
    Dexatrim for, 194t, 197, 198, 200
    ephedra-like stimulants for, 194, 194t, 197–198
    Hydroxycut for, 194t, 197–200
    patient case on, 195, 202
    Xenadrine for, 194t, 197, 198, 201
Weil, Andrew, 90, 123
whey protein supplements, 210t, 212–213, 220
white kidney bean extract, 201
willow bark, 10, 200
wine, 136t, 139–142, 141f. See also alcohol
Women's Health Initiative (WHI), 76, 91, 92
World Health Organization, on migraine, 25

X

Xenadrine, 194t, 197, 198, 201

Y

yerba mate, 196
yohimbe, 52t, 56–58, 56f
yohimbine, 56–58

Z

zinc, 174t, 179–180, 184
zinc acetate, 180
zinc gluconate, 180